Regenerative Engineering Society

Held at the 2024 AIChE Annual Meeting

San Diego, California, USA
27-31 October 2024

ISBN: 979-8-3313-1684-6

Printed from e-media with permission by:

Curran Associates, Inc.
57 Morehouse Lane
Red Hook, NY 12571

Some format issues inherent in the e-media version may also appear in this print version.

AIChE
120 Wall Street, FL 23
New York, NY 10005-4020

Phone: (800) 242-4363
Fax: (203) 775-5177

www.aiche.org

Additional copies of this publication are available from:

Curran Associates, Inc.
57 Morehouse Lane
Red Hook, NY 12571 USA
Phone: 845-758-0400
Fax: 845-758-2633
Email: curran@proceedings.com
Web: www.proceedings.com

Regenerative Engineering Society

Held at the 2024 AIChE Annual Meeting

San Diego, California, USA
27-31 October 2024

AIChE
The Global Home of Chemical Engineers

TABLE OF CONTENTS

91d Development of an Animal-Free Endothelial Cell Culture Media and Adaptation Method

Laura A.E. Brunmaier, Travis W. Walker

The successful development of a defined media for cell culture presents a host of advantages including reduction of disease transmission, facilitation of translational research, and enablement of finely tuned bioassays. Fetal bovine serum (FBS) has a nearly universal use in cell culture, but it presents consequential drawbacks, such as significant variability from batch-to-batch, serious risk of disease transmission and concerns regarding ethical collection practices. As interest continues to grow around the development of biomimetic in vitro assays and translational biomedical research, the need for full identification of components in cell media and stimulating factors have become increasingly important. Here, we present our formulation for an animal-free (AF) media and our adaptation procedures for human umbilical vein endothelial cell (HUVEC). Our goal in this study is to identify the simplest media formulation that is capable of sustaining viable HUVEC growth while optimizing for the most efficient route to adapting cells to AF conditions. This effort is intended to preserve a sufficient number of fully AF passages for experimentation. We found that coating the culture vessel with attachment proteins is critical under conditions of high or fully AF media to enhance attachment of the seeded cells and improve proliferation. We obtained conditions of cell culture that were fully AF in a chemically-defined media. This work outlines the methodologies and reasoning for implementing specific steps during the adaptation process that can be used for HUVECs and other human cell lines, and it can motivate the development of quantitative, tunable bioassays to accelerate biomedical research and translation.

91e 3D Human iPSC-Derived Immuno-Glial-Neurovascular Mibrain Tissue Model for Enhanced Modeling and Therapeutic Discovery and Development

Alice Stanton, Emre Agbas, Benjamin James, Li-Huei Tsai, Robert Langer

Up to 1 in 6 people worldwide, 1 billion people, are estimated to have a neurological disorder, costing over $500B in the US alone. For most neurological diseases there is still no pharmacologic treatment available that can slow or stop neuronal damage. To advance the discovery and development of effective treatments for neurological disease, better preclinical models are critically needed. We sought to establish a human-based, patient-specific model inclusive of integrated vascular networks and blood-brain barrier, myelinated neuronal networks, and microglia, with *in vivo*-like functionalities for enhanced preclinical studies, that can provide novel insights into disease mechanisms and effects of interventions (**Fig. 1A,B**). By engineering a novel brain-mimicking 3D hydrogel and enabling the integral co-culture of all six major brain cell types derived from patient iPSCs, we have established a 3D immuno-glial-neurovascular brain model (miBrain) (**Fig. 1C**). Taking inspiration from the human brain, we developed the Neuromatrix hydrogel to mimic the composition of brain extracellular matrix, the majority of which is comprised of glycosaminoglycans interspersed with proteoglycans surrounding neurons and glia in the interstitial space, within which are basement membrane proteins lining the vascular networks. Correspondingly, we developed a glycosaminoglycan dextran-based hydrogel with mechanical properties tuned to support the miBrain structurally, while enabling both neuronal and vascular network formation and the integral inclusion of the other four cell components, screened proteoglycan components to promote neuronal maturation and activity, and incorporated basement membrane protein mimics to facilitate vascular network formation (**Fig. 1D-I**). Differentiating each of the six major brain cell types from patient-specific iPSCs, we optimized cell protocols and developed a method for combining the cells together in Neuromatrix Hydrogel in a mixed media. This method and its underlying technologies enable a 3D immuno-glial-neurovascular co-culture inclusive of all six major brain cell types with enriched cellular phenotypes and tissue-scale phenotypes. miBrains mimic *in vivo*-like hallmarks of human brain with integrated neurovascular unit (**Fig. 1J**), enhanced *in vivo*-like microglia signatures (**Fig. 1K**), a blood-brain barrier with lumenized microvessels with tight junctions and surrounded by pericytes and astrocytic endfeet (**Fig. 1L**), myelinated neuronal networks (**Fig. 1M**), neuron maturation with robust activity and electrophysiological properties, enrichment of important *in vivo*-like transcriptional signatures, and functional interactions between cell types. This is the first brain model to our knowledge that integrates all six major brain cell types, and one which does so with biomimetic 3D tissue architecture and with all iPSC-derived cells. To model Alzheimer's Disease (AD) pathogenesis associated with the strongest genetic risk factor for sporadic AD, we constructed miBrains from cell lines of *APOE4* genetic background versus those of *APOE3*. We found that *APOE4* miBrains differentially exhibit amyloid aggregation, tau phosphorylation, and astrocytic GFAP. Harnessing the decoupled-genotype feature of our platforms, we probed the effect of *APOE4* astrocytes, identifying that *APOE4* astrocytes promote tau pathogenesis and neuronal dysregulation through crosstalk with microglia, proposing a mechanism that could be investigated for therapeutic potential. This work has established an enhanced brain tissue mimic that recapitulates human brain physiology and pathology with unprecedented biomimicry. The miBrain holds promise for enabling a new approach to target discovery for neurological disease and a valuable tool for therapeutic screening and testing of drug efficacy.

Fig. 1: Human Integrated 3D-Immuno-Glial-Neurovascular miBrain Model. (**A**) Schematic of miBrain formation harnessing patient-specific iPSCs differentiated into each of the resident brain cell types, encapsulated in Neuromatrix Hydrogel, and co-cultured for integral cell network self-assembly and microglia-like cell integration, (**B**) macroscopic view of miBrains from (i) top and (ii) side angle pictured with a dime for reference, (**C**) distribution of (from left to right) iPSC-derived pericytes (cyan: mCherry-pericytes), astrocytes (cyan: mCherry-astrocytes), BMECs (cyan: mCherry-BMECs), neurons (cyan: tubulin neuron label), oligodendroglia (cyan: tdTomato pre-transfected oligodendroglia), and iMG (cyan: membrane pre-labeled iMG) throughout the full 3D miBrain, (**D**) neuronal phenotypes in miBrains cultured in dextran-based hydrogels fabricated with various brain ECM proteins (cyan: neurofilament, blue: Hoechst; scale bars, 50 μm), (**E**) quantification of neurofilament immunoreactivity (averages of n = 8 samples from n = 3 fields of view per sample), (**F**) neuronal firing as assessed on an MEA system across conditions (n = 3 wells per group, displayed as mean and S.E.M.), (**G**) persistence of neurovascular unit phenotypes in miBrains cultured in Neuromatrix Hydrogel versus Matrigel after 5 weeks in culture (red: PECAM, cyan: TUBB3, blue: Hoechst; scale bar, 50 μm), (**H**) macroscopic view of gel structural integrity for miBrains cultured in VCN-incorporated engineered dextran-based hydrogel named Neuromatrix Hydrogel versus Matrigel after 4 weeks, (**I**) example raster plots from MEA recordings of miBrains in Neuromatrix Hydrogel compared to Matrigel; miBrains recapitulate key hallmarks of human brain tissue, inclusive of (**J**) neurovascular units: (i) 3D integrated BMEC and neuronal networks throughout the miBrain (red: Imaris reconstruction of mCherry-BMECs, cyan: tubulin neuron label, scale bar, 500 μm), (ii) visualized at higher magnification (red: Imaris surfaces of mCherry-BMECs, cyan: siR-tubulin neuron label; scale bar, 500 μm), (iii) 3D integrated astrocyte and neuronal networks throughout the miBrain (green: Imaris surfaces of mCherry-astrocytes, cyan: siR-tubulin neuron label; scale bar, 500 μm), (iv) visualized at higher magnification (green: mCherry-astrocytes, cyan: tubulin neuron label, scale bar, 100 μm), and (v) anterior view of a miBrain 3D rendering (red: PECAM, cyan: neurofilament, blue: Hoechst; scale bar, 50 μm), (**K**) microglia: (i) 3D iMG distributed throughout BMEC networks throughout the miBrain (red: Imaris surfaces from mCherry-BMECs, green: Imaris surfaces from membrane pre-labeled iMG; scale bar, 500 μm), (ii) visualized at higher magnification (red: Imaris surfaces from mCherry-BMECs, green: Imaris surfaces from membrane pre-labeled iMG; scale bar, 100 μm), (iii) 3D iMG distributed throughout neuronal networks throughout the miBrain (cyan: tubulin neuron label, green: Imaris surfaces from membrane pre-labeled iMG; scale bar, 500 μm), (iv) visualized at higher magnification (cyan: Imaris reconstruction of tubulin neuron label, green: Imaris surfaces from membrane pre-labeled iMG; scale bar, 100 μm), and (v) distribution of iMG with BMEC and neuronal networks together (cyan: tubulin neuron label, red: Imaris surfaces from mCherry-BMECs, green: Imaris surfaces from membrane pre-labeled iMG; scale bar, 100 μm), (**L**) blood-brain barrier: (i) 3D astrocytes distributed throughout BMEC networks throughout the miBrain (red: Imaris surfaces from ZO1-BMECs, green: mCherry-astrocytes; scale bar, 500 μm), (ii) visualized at higher magnification (red: Imaris surfaces from ZO1-BMECs, green: Imaris surfaces from mCherry-astrocytes; scale bar, 100 μm), (iii) further magnified (scale bar, 50 μm), (iv) ZO-1 tight junctions along vessels (green: ZO-1, red: PECAM, blue: Hoechst; scale bar, 30 μm), (v) astrocytes with end-feet extending to vessels expressing canonical aquaporin-4 transporter (green: GFAP, gray: AQP4, red: PECAM, blue: Hoechst; scale bar, 30 μm; insert, gray: AQP4), and (vi) pericytes localized to the vessels (green: NG2, red: PECAM, blue: Hoechst; scale bar, 30 μm), and (**M**) myelinated neuronal networks: (i) 3D oligodendroglia distributed throughout neuronal networks throughout the miBrain (cyan: tubulin neuron label, green: Imaris surfaces from tdTomato pre-transfected oligodendroglia; scale bar, 500 μm), (ii) visualized at higher magnification (cyan: tubulin neuron label, green: tdTomato pre-transfected oligodendroglia; scale bar, 100 μm), (iii) myelin dye-labeled neurons and oligodendroglia (red: FluoroMyelin, cyan: tubulin neuron label, green: tdTomato pre-transfected oligodendroglia, scale bar, 100 μm), (iv) visualized also via Imaris reconstructions for myelin along with neurons and oligodendroglia (red: Imaris surfaces of FluoroMyelin, cyan: tubulin neuron label, green: Imaris surfaces of tdTomato pre-transfected oligodendroglia; scale bar, 100 μm), and (v) myelin and oligodendroglia alone (red: Imaris surfaces of FluoroMyelin, green: Imaris surfaces of tdTomato pre-transfected oligodendroglia; scale bar, 100 μm), and (vi) myelination of neuronal projections (green: MBP, cyan: neurofilament, blue: Hoechst; scale bar, 10 μm).

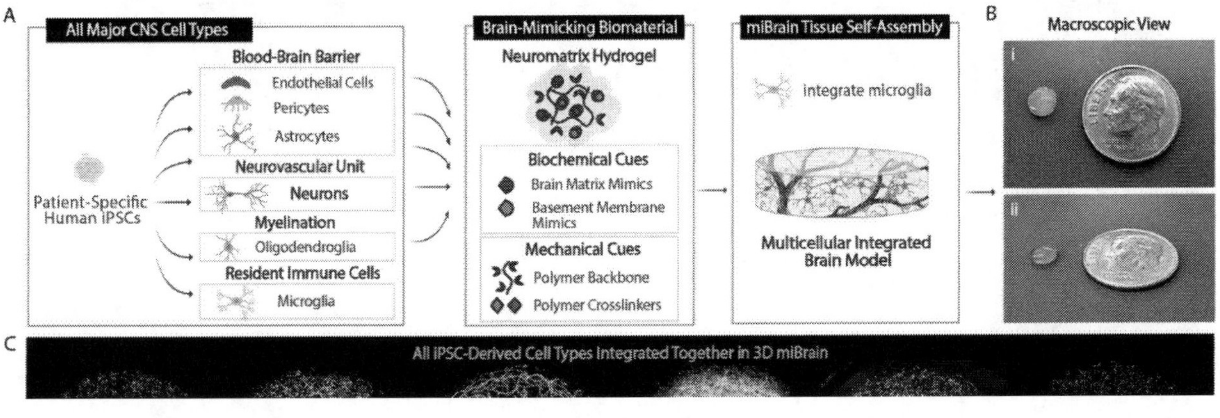

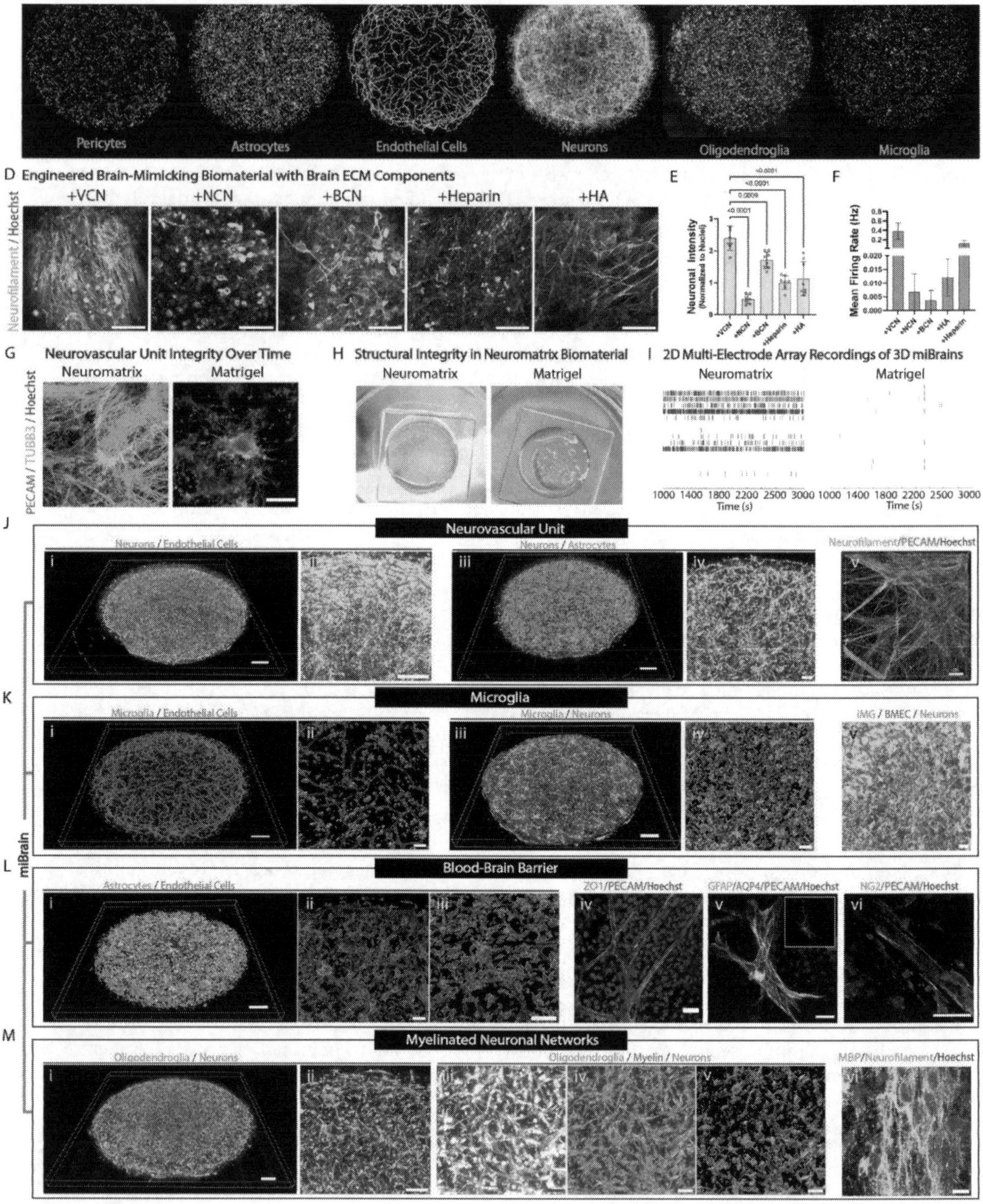
Pericytes
Astrocytes
Endothelial Cells
Neurons
Oligodendroglia
Microglia
D Engineered Brain-Mimicking Biomaterial with Brain ECM Components
Neurofilament / Hoechst
+VCN
+NCN
+BCN
+Heparin
+HA
E
Neuronal Intensity (Normalized to Nuclei)
<0.0001
0.0009
F
Mean Firing Rate (Hz)
G Neurovascular Unit Integrity Over Time
Neuromatrix
Matrigel
PECAM / TUBB3 / Hoechst
H Structural Integrity in Neuromatrix Biomaterial
Neuromatrix
Matrigel
I 2D Multi-Electrode Array Recordings of 3D miBrains
Neuromatrix
Matrigel
1000 1400 1800 2200 2500 3000
Time (s)
J
Neurovascular Unit
Neurons / Endothelial Cells
Neurons / Astrocytes
Neurofilament/PECAM/Hoechst
K
Microglia
Microglia / Endothelial Cells
Microglia / Neurons
iMG / BMEC / Neurons
miBrain
L
Blood-Brain Barrier
Astrocytes / Endothelial Cells
ZO1/PECAM/Hoechst
GFAP/AQP4/PECAM/Hoechst
NG2/PECAM/Hoechst
M
Myelinated Neuronal Networks
Oligodendroglia / Neurons
Oligodendroglia / Myelin / Neurons
MBP/Neurofilament/Hoechst

Bioactive Materials 27 (2023) 168–180

Contents lists available at ScienceDirect

KeAi
CHINESE ROOTS
GLOBAL IMPACT

Bioactive Materials

journal homepage: www.keaipublishing.com/en/journals/bioactive-materials

Engineered human pluripotent stem cell-derived natural killer cells with PD-L1 responsive immunological memory for enhanced immunotherapeutic efficacy

Yun Chang [a,b], Gyuhyung Jin [a,b], Weichuan Luo [d], Qian Luo [d], Juhyung Jung [a], Sydney N. Hummel [a], Sandra Torregrosa-Allen [b], Bennett D. Elzey [b,c], Philip S. Low [b,d,**], Xiaojun Lance Lian [e,***], Xiaoping Bao [a,b,*]

[a] Davidson School of Chemical Engineering, Purdue University, West Lafayette, IN, 47907, USA
[b] Purdue University Institute for Cancer Research, West Lafayette, IN, 47907, USA
[c] Department of Comparative Pathobiology, Purdue University, West Lafayette, IN 47907, USA
[d] Department of Chemistry, Purdue University, West Lafayette, IN, 47907, USA
[e] Department of Biomedical Engineering, The Huck Institutes of the Life Sciences, Department of Biology, The Pennsylvania State University, University Park, PA, 16802, USA

A R T I C L E I N F O

Keywords:
Natural killer cells
Human pluripotent stem cells
Immunological memory
Immunotherapy
Chimeric antigen receptor

A B S T R A C T

Adoptive chimeric antigen receptor (CAR)-engineered natural killer (NK) cells have shown promise in treating various cancers. However, limited immunological memory and access to sufficient numbers of allogenic donor cells have hindered their broader preclinical and clinical applications. Here, we first assess eight different CAR constructs that use an anti-PD-L1 nanobody and/or universal anti-fluorescein (FITC) single-chain variable fragment (scFv) to enhance antigen-specific proliferation and anti-tumor cytotoxicity of NK-92 cells against heterogenous solid tumors. We next genetically engineer human pluripotent stem cells (hPSCs) with optimized CARs and differentiate them into functional dual CAR-NK cells. The tumor microenvironment responsive anti-PD-L1 CAR effectively promoted hPSC-NK cell proliferation and cytotoxicity through antigen-dependent activation of phosphorylated STAT3 (pSTAT3) and pSTAT5 signaling pathways via an intracellular truncated IL-2 receptor β-chain (ΔIL-2Rβ) and STAT3-binding tyrosine-X-X-glutamine (YXXQ) motif. Anti-tumor activities of PD-L1-induced memory-like hPSC-NK cells were further boosted by administering a FITC-folate bi-specific adapter that bridges between a programmable anti-FITC CAR and folate receptor alpha-expressing breast tumor cells. Collectively, our hPSC CAR-NK engineering platform is modular and could constitute a realistic strategy to manufacture off-the-shelf CAR-NK cells with immunological memory-like phenotype for targeted immunotherapy.

1. Introduction

Adoptive cellular therapy (ACT)-based immunotherapies have emerged as powerful and potentially curative therapies for the treatment of various types of cancers [1–3]. Natural killer (NK) cell-based immunotherapies [4–6] are increasingly attractive due to their innate anti-tumor immunity without prior sensitization and/or antigen presentation [7,8]. Furthermore, allogenic NK cells do not induce graft-versus-host disease (GvHD) after infusion, which is commonly associated with allogenic T cell transplantation [9,10]. While significant clinical benefit has been achieved in the treatment of hematologic malignancies and melanoma, the efficacy of adoptive NK cell therapies in treating solid tumors is thus far limited due in part to the failure of transferred NK cells to develop classical immunological memory [11,

Peer review under responsibility of KeAi Communications Co., Ltd.
* Corresponding author. Davidson School of Chemical Engineering, Purdue University, West Lafayette, IN, 47907, USA.
** Corresponding author. Purdue University Center for Cancer Research, West Lafayette, IN, 47907, USA.
*** Corresponding author.
E-mail addresses: plow@purdue.edu (P.S. Low), lian@psu.edu (X.L. Lian), bao61@purdue.edu (X. Bao).

https://doi.org/10.1016/j.bioactmat.2023.03.018
Received 1 December 2022; Received in revised form 14 February 2023; Accepted 22 March 2023
Available online 5 April 2023

12], which is mainly caused by the inability of receptor genes in NK cells to undergo rearrangement and the exhaustion of NK cells exposed to an immunosuppressive tumor microenvironment [13,14]. Thus, engineering NK cells with tumor microenvironment responsive chimeric antigen receptors (CARs) holds great promise for achieving immunological memory-like activities of NK cells during tumor ablation.

Specific receptor stimulation promotes significant expansion of NK cells in a diseased microenvironment, and these self-renewal memory NK cells rapidly degranulate and produce cytokines upon reactivation to perform robust protective immunity [15,16]. While various tumor targeting CARs in NK cells can be effectively stimulated by specific tumor antigens, their therapeutic efficacies can still be limited due to the poor *in vivo* expansion and persistence of NK cells after infusion. The employment of cytokines, such as IL-15 [17–19], IL-18 [10,16,20–22], and IL-21 [23–25], has been widely reported to enhance *in vivo* persistence and/or memory of various NK cells. However, cytokine stimulation may lead to autonomous NK cell proliferation, severe adverse events, or even leukemia transformation in patients [26]. To produce superior memory-like NK cells in a robust, safe and controllable way, CAR structures should be designed to effectively and specifically recognize immunosuppressive signals in the tumor microenvironment and immediately activate intracellular proliferation signaling pathways in NK cells [18,19], leading to tumor-responsive cellular expansion and prevention of NK cell exhaustion [27]. Among these immunosuppressive signals, programmed death-ligand 1 (PD-L1), which is expressed on various solid tumor cells and interact with PD-1 on immune cells to suppress immunotherapy [28–30], are widely used in CAR design, since PD-L1/PD-1 blockade has achieved significant clinical benefits [31,32]. In addition to enhanced *in vivo* persistence, these memory-like NK cells should have excellent tumor-killing ability. CAR constructs containing transmembrane and/or co-stimulatory domains of NKG2D, 2B4, and 41BB were reported to effectively activate intracellular cytotoxicity signaling pathways in NK cells [33], but continuous exposure to antigens may cause NK cell exhaustion and prevent acquisition of a memory-like phenotype [34,35]. To avoid T cell exhaustion and cytokine storm [36–39], an anti-fluorescein (FITC) single-chain variable fragment (scFv)-based CAR has been used in T cells to eradicate tumor cells only in the presence of a low molecular weight adapter [40]. These fluorescein-cancer bridging small molecules have a short circulation half-life <90 min and can easily penetrate solid tumors [39]. Such a bi-specific adapter strategy may also be used to prevent NK cell exhaustion, enhance broader applicability against heterogenous solid tumors, and reduce off-target toxicity to the non-target organs.

Broader application of adoptive NK cell therapies is currently hindered by the limited access to sufficient numbers of donor cells for multiple-dose transplantations [41]. Additionally, genetic modification of primary NK cells is technically challenging and laborious, leading to heterogeneous CAR-NK cells [42]. In contrast, human pluripotent stem cells (hPSCs) are capable of self-renewal and can be genetically modified to produce customized NK cells in a homogenous and clinically-scalable manner [7,33,43], rendering them as a promising platform for making realistically off-the-shelf cellular immunotherapies.

In this study, we first assessed eight different CAR constructs that use an anti-PD-L1 nanobody and/or universal anti-FITC scFv to enhance antigen-specific proliferation and anti-tumor cytotoxicity of NK cells (Fig. 1). hPSCs were then engineered with these optimized CARs and differentiated into functional dual CAR-NK cells. The tumor microenvironment responsive anti-PD-L1 CAR effectively promoted hPSC-NK cell proliferation and cytotoxicity against tumor cells through antigen-dependent activation of phosphorylated STAT3 (pSTAT3) and pSTAT5 signaling via an intracellular truncated IL-2 receptor β-chain (ΔIL-2Rβ) and a STAT3-binding tyrosine-X-X-glutamine (YXXQ) motif. Anti-tumor activities of PD-L1-induced memory-like hPSC-NK cells were further boosted by treatment with a FITC-folate adapter that bridges between the programmable anti-FITC CAR and folate receptor alpha (FRα)-expressing breast tumor cells, but not the FRα rare prostate cancer cells.

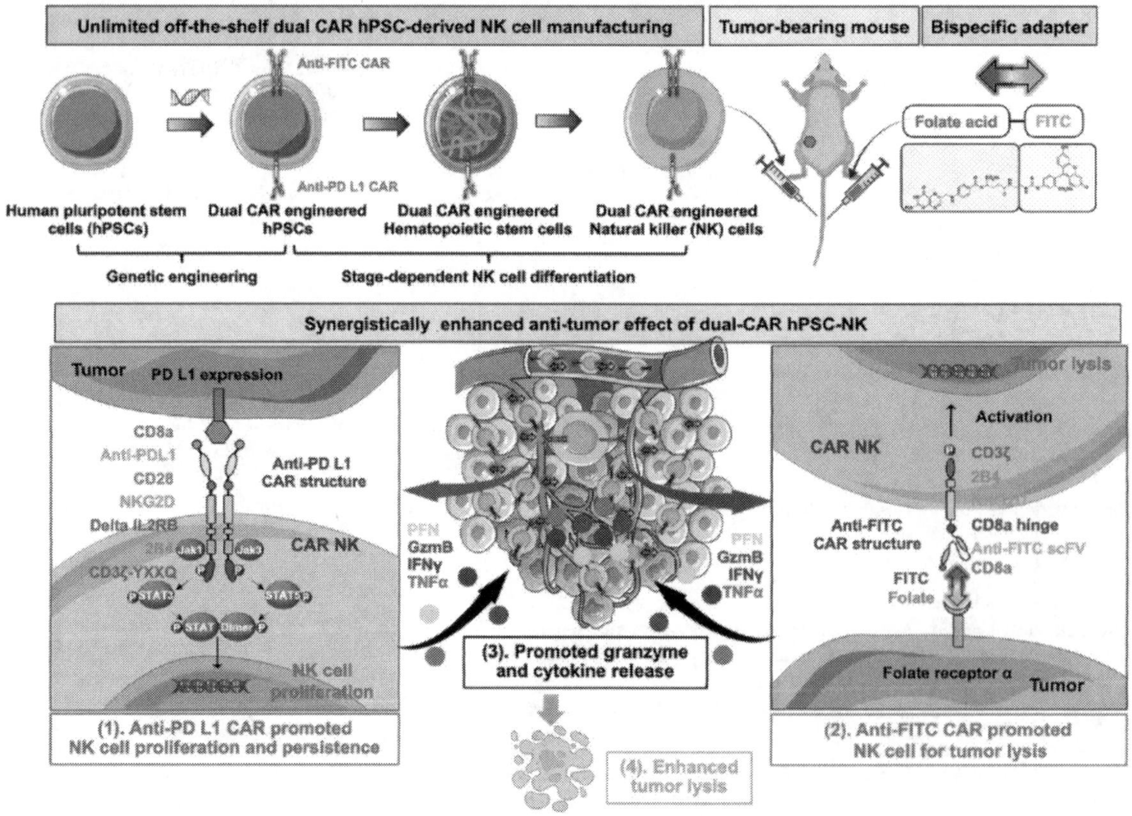

Fig. 1. Schematic of synergistically enhanced anti-tumor effect of dual CAR hPSC-NK cells.

Notably, anti-FITC CAR is also switchable to target other tumor cells, including LNCaP using a bi-specific adapter with affinity for PSMA, namely FITC-DUPA. Collectively, our hPSC CAR-NK engineering platform is designed to be modular and adaptable for large scale manufacturing of off-the-shelf CAR-NK cells with immunological memory-like phenotype for targeted immunotherapy.

2. Results

2.1. Screening CAR structures with enhanced NK cell-mediated tumor-killing activities

Based on previous CAR constructs used in T and NK cells [33,39], we first designed and evaluated 8 different CAR designs optimized for anti-tumor cytotoxicity and proliferation in NK-92 cells (Fig. 2A). CAR #1 to #4 were single antigen-targeting CARs against either PD-L1 or FITC using NK or T cell-specific signaling domains, and CAR #5 to #8 were combinatory dual antigen-targeting CARs. For the switchable anti-FITC scFv CARs, CAR #1, #5, and #7 employ a NK-specific transmembrane domain NKG2D, a co-stimulatory domain 2B4 and an intracellular domain CD3ζ, whereas CAR #2, #6, and #8 differ in the transmembrane domain CD8 and co-stimulatory domain 4-1BB. For tumor microenvironment responsive anti-PD-L1 nanobody CARs, CAR #3, #5, and #7 use a NK-specific transmembrane domain NKG2D, a co-stimulatory domain 2B4, a truncated IL-2 receptor β-chain (Delta IL-2RB), an intracellular domain CD3ζ, and a STAT3-binding

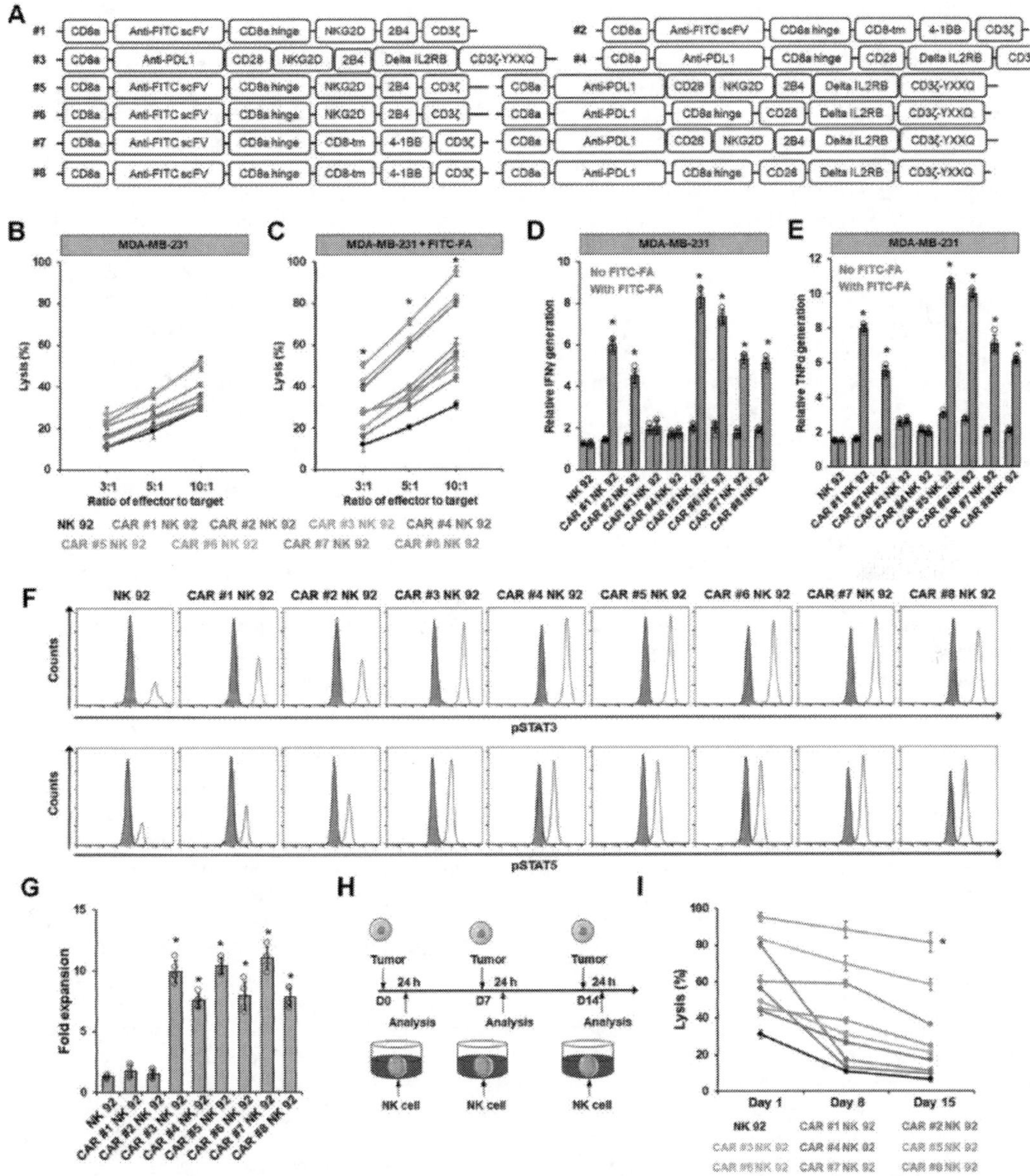

Fig. 2. Screening CAR structures with enhanced anti-tumor activity and proliferation in NK-92 cells. (A) Schematic of various lentiviral CAR constructs. Killing of MDA-MB-231 tumor cells by indicated NK-92 cells was performed at different ratios of effector-to-target in the absence (B) or presence (C) of 10 nM FITC-folate adapter. Data are represented as mean ± s.d. of five independent replicates, $*p < 0.05$. ELISA analysis of secreted cytokine IFNγ (D) and TNFα (E) from various NK-92 cells upon MDA-MB-231 stimulation was shown. Data are represented as mean ± s.d. of five independent replicates, $*p < 0.05$. (F) Representative flow cytometry analysis of phosphorylated STAT3 (pSTAT3) and STAT5 (pSTAT5) in indicated NK-92 cells upon MDA-MB-231 stimulation. (G) Expansion of indicated NK-92 cells 7 days after co-culture with MDA-MB-231 cells was quantified. Data are represented as mean ± s.d. of five independent replicates, $*p < 0.05$. (H-I) Schematic of *in vitro* MDA-MB-231 tumor rechallenge model and cytotoxicity assay was shown in (H). (I) Killing of MDA-MB-231 tumor cells by indicated NK-92 cells was performed in the presence of 10 nM FITC-folate adapter at different time points. Data are represented as mean ± s.d. of five independent replicates, $*p < 0.05$.

tyrosine-X-X-glutamine (YXXQ) motif, whereas CAR #4, #6, and #8 differ in the transmembrane domain CD28. These CAR constructs were first tested in NK-92 cells for their ability to enhance anti-tumor activities against FRα+ and PD-L1+ tumor cells. Human breast cancer MDA-MB-231 cells express high levels of FRα and PD-L1, whereas human prostate adenocarcinoma LNCaP cells express neither FRα nor PD-L1 (Fig. S1A) [44]. These two tumor lines were used for anti-tumor cytotoxicity analysis of our engineered CAR NK-92 cells.

Bi-specific FITC-folate adapter (FITC-FA) was first synthesized with folic acid on its left resides for binding FRα on breast tumor cells and fluorescein on the right side for anti-FITC CAR targeting (Fig. S1B) [39]. The binding affinity (K_d) of FITC-folate for MDA-MB-231 tumor cells was measured as 2.64 nM (Fig. S1C), and the affinity (K_d) of FITC-folate for various anti-FITC CAR NK-92 cells were about 10 nM (Fig. S1D). Considering the fact that insufficient intracellular bridges will be formed at very low FITC-folate concentration, whereas at very high concentrations, intracellular bridging will be locked due to monovalent saturation of ligand binding sites on both cell types with excess FITC-folate adapters, 10 nM of FITC-folate was used in the following studies.

The killing potency of various anti-FITC and/or anti-PD-L1 CAR NK-92 cells (Fig. S1E) were then tested in MDA-MB-231 (FRα+ PD-L1+) and LNCaP cell (FRα- PD-L1-) cells. As expected, CAR-expressing NK-92 cells exhibited more potent cytotoxicity against MDA-MB-231 and more cytotoxic granule release than LNCaP cells (Fig. 2B and **S2A-D**). In the presence of bi-specific FITC-folate adapter, anti-MDA-MB-231 cytotoxicity of CAR NK-92 cells was significantly increased (Fig. 2C), indicating the specificity of our anti-FITC CAR. Among these CARs, CAR #1, #5, and #6 displayed a much larger increase of anti-tumor activity in NK-92 cells against FRα+ PD-L1+ breast cancer cells after bridging with the FITC-folate adapter (Fig. 2C), along with significantly enhanced IFNγ and TNFα release (cytotoxic granule) (Fig. 2D–E). As expected, these FITC-folate bridged NK-CARs (#1, #5, and #6) mediated higher killing potency in NK-92 cells than T cell specific CARs (#2, #7, and #8). To further test the broader applicability of our anti-FITC CAR-NK cells in treating heterogenous solid tumors, we also synthesized a bi-specific FITC-DUPA adapter (Fig. S3A-**B**) to target prostate specific membrane antigen (PSMA)-expressing prostate tumor cells. As expected, FITC-DUPA-bridged CAR #1, #5, and #6 NK-92 cells lysed more LNCaP cells than MDA-MB-231 cells (Figs. S3C–S3D), confirming the specificity and multi-tumor targeting ability of our anti-FITC CAR.

2.2. Screening CAR structures with enhanced NK cell proliferation activity

We next evaluated the capability of various CARs in promoting antigen-specific NK cell proliferation after co-culturing with tumor cells. Both truncated IL-2 receptor β-chain (ΔIL-2RB) and the STAT3-binding tyrosine-X-X-glutamine (YXXQ) motif in the anti-PD-L1 CARs were designed to enhance cell proliferation and persistence via activation of JAK, STAT3 and STAT5 signaling pathways [45]. Upon PD-L1+ MDA-MB-231 cell stimulation, CAR NK-92 cells exhibited upregulated levels of phosphorylated STAT3 (pSTAT3) and pSTAT5 (Fig. 2F), among which CAR #3, #5, and #7 displayed superior ability in upregulating pSTAT3 and pSTAT5 (Figs. S4A–S4B). As expected, CAR #3, #5, and #7 also promoted greatest proliferation in NK-92 cells (Fig. 2G). We next constructed a continuous *in vitro* tumor cell exposure model to investigate the persistence and memory-like phenotype of NK-92 cells after CAR engineering (Fig. 2H). Consistent with previous observation, NK-92 cells engineered with CAR #1, #5 and #6 displayed superior tumor-killing ability against FRα+ PD-L1+ breast cancer cells under the initial antigen exposure at day 1 (Fig. 2I). As the antigen exposure increases (day 8 and 15), significant reduction of anti-tumor cytotoxicity in NK-92 cells with anti-FITC CAR only (CAR #1) was observed, whereas dual anti-FITC and anti-PD-L1 CAR NK-92 cells (CAR #5 and #6) still exhibited excellent anti-tumor activities at day 15. Notably, dual anti-FITC and anti-PD-L1 CAR #5 with NK-specific transmembrane and co-stimulatory domains presented superior persistence as compared to all other CARs. Recently, memory-like NK cells were induced with a cytokine cocktail of IL-12, IL-15 and IL-18 [15,16]. We next evaluated if cytokine treatment may achieve boosted memory-like phenotype in our dual CAR-NK cells (Fig. S5A). While cytokine treatment increased anti-tumor cytotoxicity of single anti-PD-L1 CAR NK-92 cells, significant difference was not observed in FITC-FA bridged dual CAR NK-92 cells (Fig. S5B). Our results demonstrated that dual CAR design may synergize multi-functions of NK cells under specific tumor antigen stimulation and achieve a memory-like phenotype with superior anti-tumor activities and persistence under immunosuppressive tumor microenvironment.

2.3. Engineering hPSC-derived NK cells with dual CARs for enhanced function

Given its superior anti-tumor activity and persistence in NK-92 cells, dual anti-FITC and PD-L1 CAR #5 was selected for CAR engineering of hPSC-derived NK cells. Single antigen-targeting anti-FITC CAR #1 and anti-PD-L1 CAR #3 was used as a control for anti-tumor cytotoxicity and cell proliferation, respectively. To provide a potentially off-the-shelf source of CAR-expressing NK cells, we engineered hPSCs with these three CARs. We knocked the universal anti-FITC CAR into the *AAVS1* safe harbor locus via CRISPR/Cas9-mediated homologous recombination [46] (Figs. S6A–B), leading to robust CAR-expressing hPSCs (Fig. S6C). Notably, engineered hPSCs retained high expression levels of pluripotency markers, including stage-specific embryonic antigen-4 (SSEA-4) and octamer-binding transcription factor 4 (OCT-4) (Fig. S6C). To determine the effect of CAR expression on NK cell differentiation, genetically-modified hPSCs were subjected for hematopoietic [47] and NK cell differentiation [48] using stage-specific morphogens (Fig. S7A). High purity of CD45$^+$CD43$^+$ hematopoietic stem and progenitor cells (HSPCs) (Fig. S7B) and CD56$^+$CD45$^+$ NK cells (Fig. S7C) were successfully generated from wildtype or CAR-expressing hPSCs. The resulting hPSC-derived NK cells also expressed high levels of typical NK cell surface markers, including CD16, KID3DL1, NKp46, NKG2D, and NKp44 (Fig. 3A).

To determine their anti-tumor cytotoxicity, CAR-expressing hPSC-derived NK cells were co-cultured with MDA-MB-231 cells in the presence of 10 nM FITC-folate. As compared to wildtype hPSC-NK cells, more immunological synapses were formed between CAR-engineered NK cells within 2 h (Fig. 3B), and dual CAR-NK cells formed most immunological synapses with tumor cells (Fig. 3C), whereas all hPSC-derived NK cells showed a similar but weaker immunological synapse formation ability against FRα-PD L1- LNCaP prostate cancer cells (Fig. S8A), demonstrating the high specificity of these CARs to the targeted tumor antigens. In response to MDA-MB-231 tumor cells, CAR-NK cells expressed more IFNγ and CD107a (Fig. 3D) and released more cytotoxic granule TNFα and IFNγ (Fig. 3E–F). As expected, dual CAR-NK cells expressed most IFNγ and CD107a, and released the most cytotoxic granule, whereas all tested NK cells expressed low levels of IFNγ and CD107a, and released low amount of cytotoxic granule upon FRα-PD-L1- LNCaP cell stimulation (Fig. S8B-**D**). We next assessed the tumor-killing ability of different hPSC-NK cells, and demonstrated that dual CAR-NK cells displayed superior anti-MDA-MB-231 cytotoxicity as compared to wildtype, anti-FITC CAR, and anti-PD-L1 CAR-NK cells (Fig. 3G), whereas all hPSC-derived NK cells displayed similar and low cytotoxicity against LNCaP tumor cells (Fig. S8E).

We then investigated the antigen-responsive proliferation ability of various hPSC-NK cells. Upon PD-L1+ MDA-MB-231 cell stimulation, hPSC-derived CAR-NK cells upregulated expression levels of phosphorylated STAT3 (pSTAT3) and pSTAT5 (Fig. S9A). Single antigen-targeting anti-PD-L1 and dual CAR-NK cells exhibited highest expression levels of pSTAT3 and pSTAT5 (Fig. 3H), and achieved highest cell expansion (Fig. 3I). We next investigated the anti-tumor cytotoxicity and persistence of CAR-NK cells in a continuous antigen exposure model. While

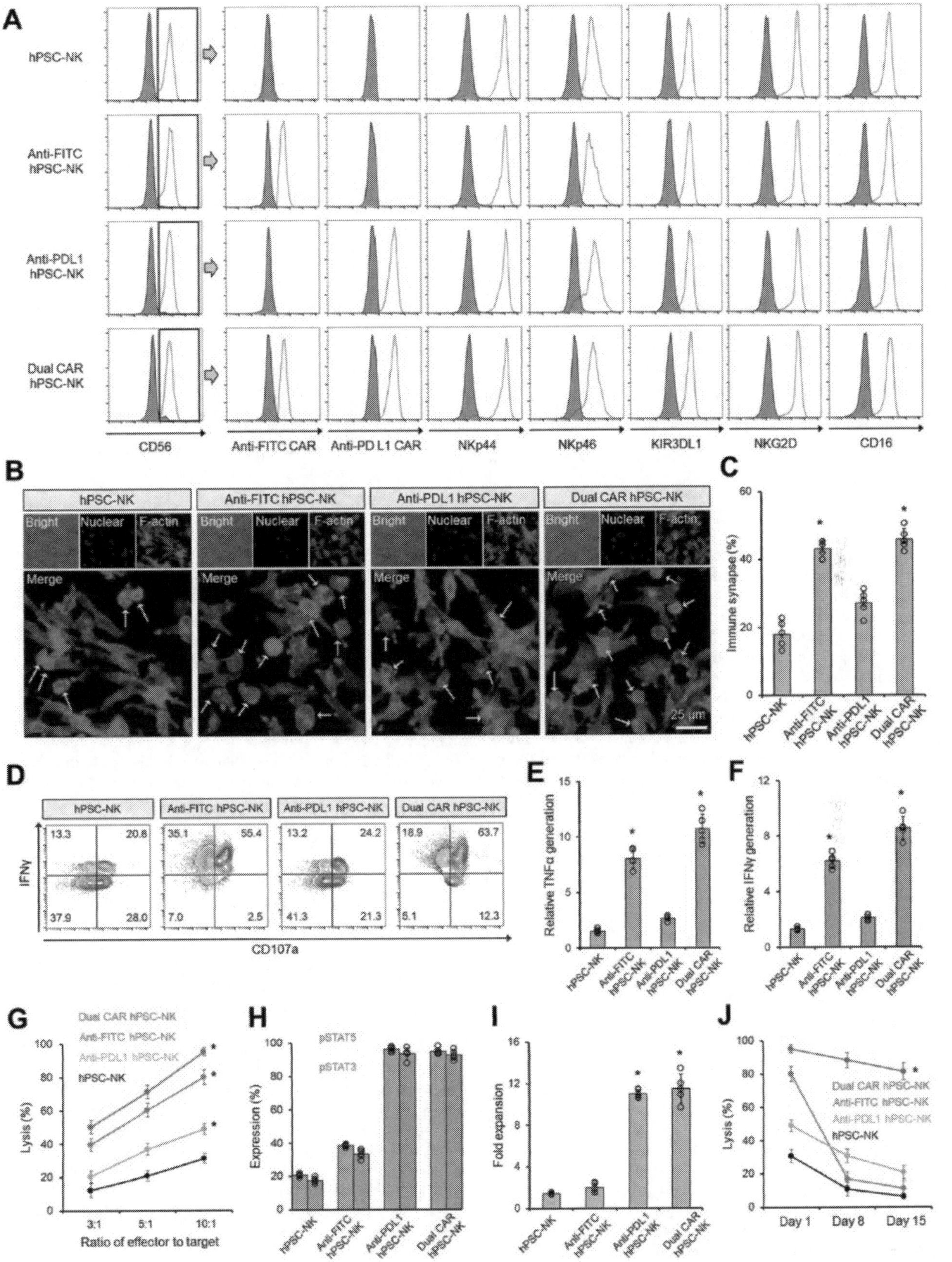

(caption on next page)

Fig. 3. NK cells derived from chimeric antigen receptor (CAR)-engineered H9 hPSCs display enhanced memory-like anti-tumor cytotoxicity. (A) Flow cytometry analysis of NK cells derived from different hPSCs. Plots show histograms of control (Black) and indicated NK cell-specific antibody (blue). (B-C) Representative images of immunological synapses at the interface between tumor and indicated hPSC-NK cells by F-actin staining (B) and the numbers of immunological synapses were quantified in (C). Scale bar, 25 μm. (D) Flow cytometry analysis of INFγ/CD107a in different NK cells upon MDA-MB-231 cell stimulation. ELISA analysis of secreted cytokine TNFα (E) and INFγ (F) from indicated NK cells in response to MDA-MB-231 cells is shown. (G) Killing of MDA-MB-231 tumor cells by indicated hPSC-NK cells was performed at different effector-to-target ratios in the presence of 10 nM FITC-folate adapter. (H) Expression of phosphorylated STAT3 (pSTAT3) and STAT5 (pSTAT5) in the indicated hPSC-NK cells upon MDA-MB-231 stimulation was quantified. (I) Expansion of indicated hPSC-NK cells 7 days after co-culture with MDA-MB-231 tumor cells was quantified. (J) Killing of MDA-MB-231 tumor cells by indicated hPSC-NK cells was performed in the presence of 10 nM FITC-folate adapter at different time points. Data are represented as mean ± s.d. of five independent replicates, $^{*}p < 0.05$.

similar strong initial anti-MDA-MB-231 cytotoxicity was observed in anti-FITC and dual CAR NK cells at day 1 (Fig. 3J), anti-FITC CAR-NK cells significantly reduced tumor-killing ability as antigen exposure time increase (day 8 and 15), whereas dual CAR-NK cells still exhibited excellent anti-tumor ability and persistence at day 15. Consistent with observation in NK-92 cells, memory-like NK cytokine treatment significantly enhanced anti-tumor cytotoxicity of hPSC-derived wildtype or single CAR-NK cells (Figs. S9B–C). Significant difference was not observed between FITC-FA bridged dual CAR-NK cells treated with or without cytokine, indicating the memory-like potency of our dual CAR. Importantly, all CAR-expressing hPSC-derived NK cells did not kill normal H9 hPSCs and hPSCs-derived somatic cells (Fig. S9D), demonstrating their safety in future clinical applications.

2.4. Dual CAR-hPSC-NK cells have improved antigen-responsive persistence in vivo

Systemic administration or ectopic expression of interleukin-15 (IL-15) has been used to improve *in vivo* persistence of CAR-NK cells [17–19], which though may lead to abnormal cell proliferation, severe adverse events or even leukemia transformation in patients [26]. To determine the effect of anti-PD-L1 CAR on the proliferation and persistence of hPSC-NK cells, immunodeficient NRG mice engrafted with 5×10^5 PD-L1-expressing MDA-MB-231 breast cancer cells or PD-L1-rare LNCaP cells (Fig. 4A, Figs. S10-11) were treated with intravenous infusion of 5×10^6 different hPSC-derived NK cells or PBS 7 days after tumor cell injection. These CAR NK cells exhibited excellent tumor targeting ability (Fig. S10). Host blood was collected at day 6, 14, 21, and 28 for NK cell analysis, and significantly higher NK cell numbers were detected in the anti-PD-L1 and dual CAR NK groups in the MDA-MB-231 mouse xenograft tumor model than other groups (Fig. 4B–C). As expected, low NK cells were detected in all experimental groups of LNCaP xenograft model (Fig. S11A-**B**), highlighting the specificity of anti-PD-L1 CAR and its capacity to enhance persistence of NK cells *in vivo*.

The biocompatibility of hPSC-derived CAR-NK cells was also evaluated by monitoring the body weight of host mice, and there was no significant body weight loss across all tested experimental groups (Fig. 4D, **S10C**), indicating the minimal systemic toxicity and high biocompatibility of hPSC-derived NK cells. Histological analysis of major organs sliced from host mice at day 30 showed that adoptive NK cells did not cause any observable abnormality or damage in heart, liver, spleen, lung, and kidney (Fig. 4E), which was further supported by the blood biochemistry analysis (Table S1), confirming the biocompatibility of our hPSC-derived NK cells.

2.5. Dual CAR-hPSC-NK cells have improved anti-tumor activities in tumor rechallenge models

To evaluate anti-tumor activities of different hPSC-NK cells *in vivo*, we first tested cytotoxicity of MDA-MB-231 cells in a mouse xenograft model. Mice were subcutaneously injected with 5×10^5 MDA-MB-231 cells and systemically administered a single injection of 1×10^7 NK cells 7 days after tumor inoculation (Fig. 5A). As compared to the tumor-bearing mice treated with PBS, administration of hPSC-NK cells significantly reduced tumor burden (Fig. 5B–D). As expected, dual CAR hPSC-NK cells displayed higher anti-tumor cytotoxicity than wildtype or other CAR-expressing NK cells. We next measured human cytokine production release in the plasma of different experimental mouse groups, including TNFα, IL-6 and IFNγ. All non-PBS experimental groups released detectable TNFα and IL-6 in the plasma from day 14 to day 28, and dual CAR hPSC-NK cells maintained highest levels of both cytokines (Fig. 5E and F), which were eventually decreased in host mice, indicating a reduced risk of cytokine release syndrome. The superior production of IFNγ in dual CAR-NK cells (Fig. 5G) is also consistent with their memory-like phenotype [15,16].

Given the promising *in vivo* performance of our hPSC-derived NK cells, MDA-MB-231 tumor cells were re-inoculated to construct a tumor rechallenge model for the investigation of their memory-like behavior (Fig. 6A). Compared with other experimental groups, dual CAR hPSC-NK cells significantly slowed down the growth of second tumor (Fig. 6B–C and Fig. S12) and prolonged the survival of tumor-bearing mice (Fig. 6D). Our data indicate that the combination of tumor microenvironment responsive anti-PD-L1 and programmable anti-FITC CARs significantly enhances *in vivo* persistence and anti-tumor activities of hPSC-derived NK cells, endowing them with a memory-like capacity for improved immunotherapy.

3. Discussion

Due to their innate immunity against all kinds of pathogen and unique property of not causing GvHD in allogenic transplantation, adoptive CAR-NK cell therapy holds great promise in treating various cancers. To date, most efforts have been conducted to treat hematologic malignancies with FDA-approved anti-CD19 or anti-CD33 CAR NK cells [18,49–51]. CARs targeting various solid tumors, such as EGFR$^+$ breast cancer [52,53], EGFR$^+$/HER2$^+$ glioblastoma [17,54], and PSMA$^+$ prostate cancer [55], are also developed for preclinical and clinical studies. While none of the 11 patients in a phase I/II trial developed severe toxicity [51], concerns still exist regarding the off-tumor effects, sustained effector cell proliferation and exhaustion, cytokine release syndrome (CRS), and CAR specificity against targeted tumor, particularly in antigen-heterogeneous cancers. To circumvent these concerns in CAR-T cells [56], we have previously developed a universal anti-FITC CAR strategy, in which cancer cell killing can be precisely controlled by regulating doses of a bispecific adapter and CAR expression on T cells can be sensitively regulated by a fluorescein-linked TLR7 to selectively activate CAR-T cells and sensitively manage a CRS [36,39]. Such concerns may also be observed in adoptive CAR-NK cell therapies, posing a significant barrier for their clinical translation. To develop a safer CAR-NK cell therapy that are truly universal and potent, we equipped NK cells with a stable anti-FITC CAR, and demonstrated the controllable and potent anti-tumor activities of hPSC-derived CAR-NK and CAR-NK-92 cells, though a comprehensive safety profile of our CAR-NK cells is still needed in future preclinical and clinical studies.

In the pursuit of achieving higher efficacy and potency in NK cells, we also added a second anti-PD-L1 CAR in our CAR-NK cells due to the striking clinical efficacy shown by checkpoint inhibitors that target PD-1 or PD-L1. Targeting PD-L1 may allow selective targeting of solid tumor cells and side effect profiles would be predicted based on PD-1/L1 immune checkpoint blockade [57]. Clinical trials with anti-PD-L1 CAR-NK cells are also currently underway (NCT04847466). Unlike previous

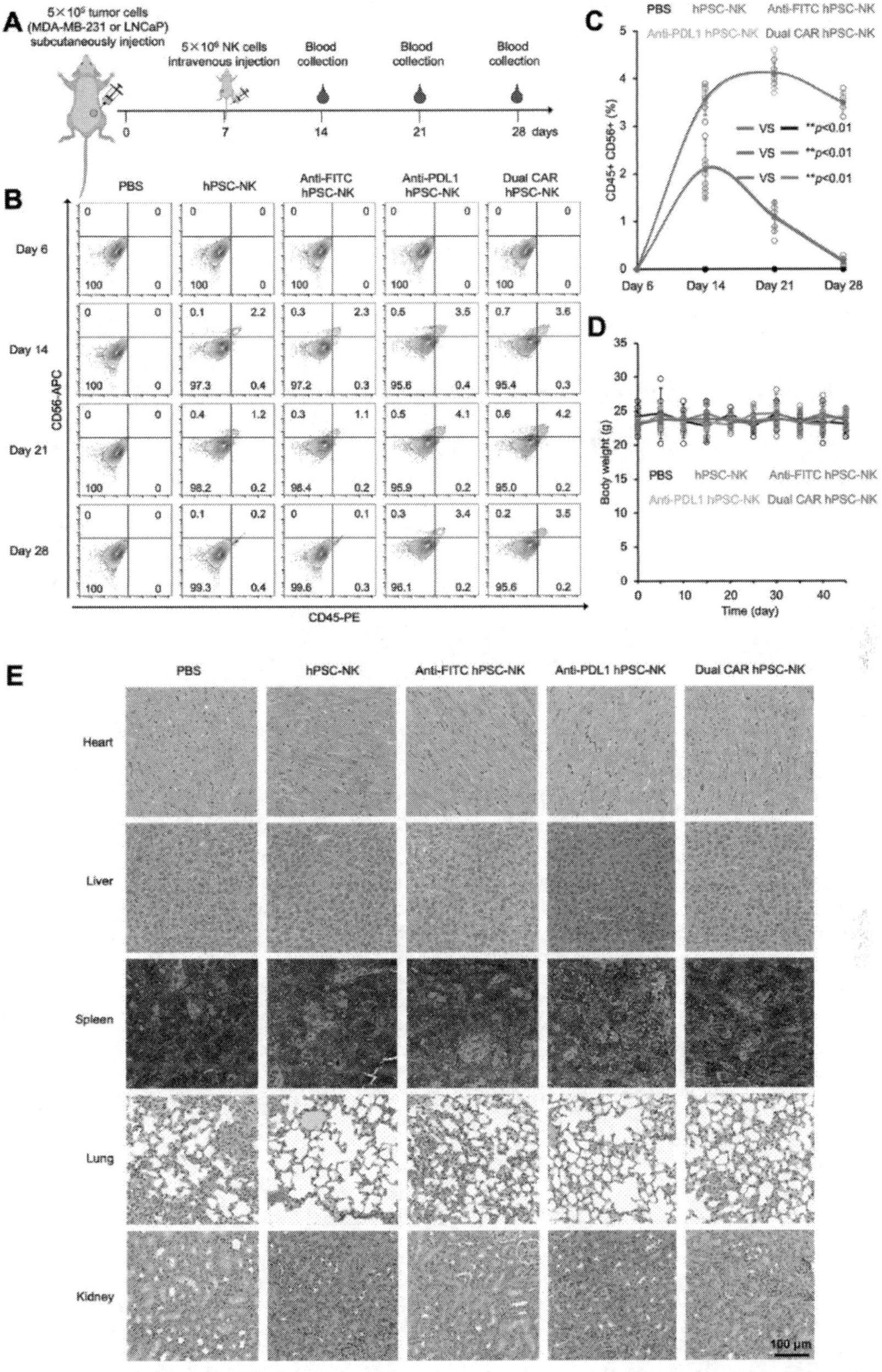

Fig. 4. Dual CAR-hPSC-NK cells have improved persistence *in vivo*. (A) Schematic of subcutaneous injection of MDA-MB-231 cells for *in vivo* tumor model construction and persistence analysis of various hPSC-derived NK cells. (B) Flow cytometry analysis of $CD45^+CD56^+$ hPSC-NK cells in host blood at different time points after intravenous injection of indicated hPSC-NK cells or PBS control. The percentage of $CD45^+CD56^+$ hPSC-NK cells was quantified in (C). n = 5. (D) Body weight of all experimental mouse groups was measured at indicated time points. (E) H&E images of major organs collected at the end of treatment are shown.

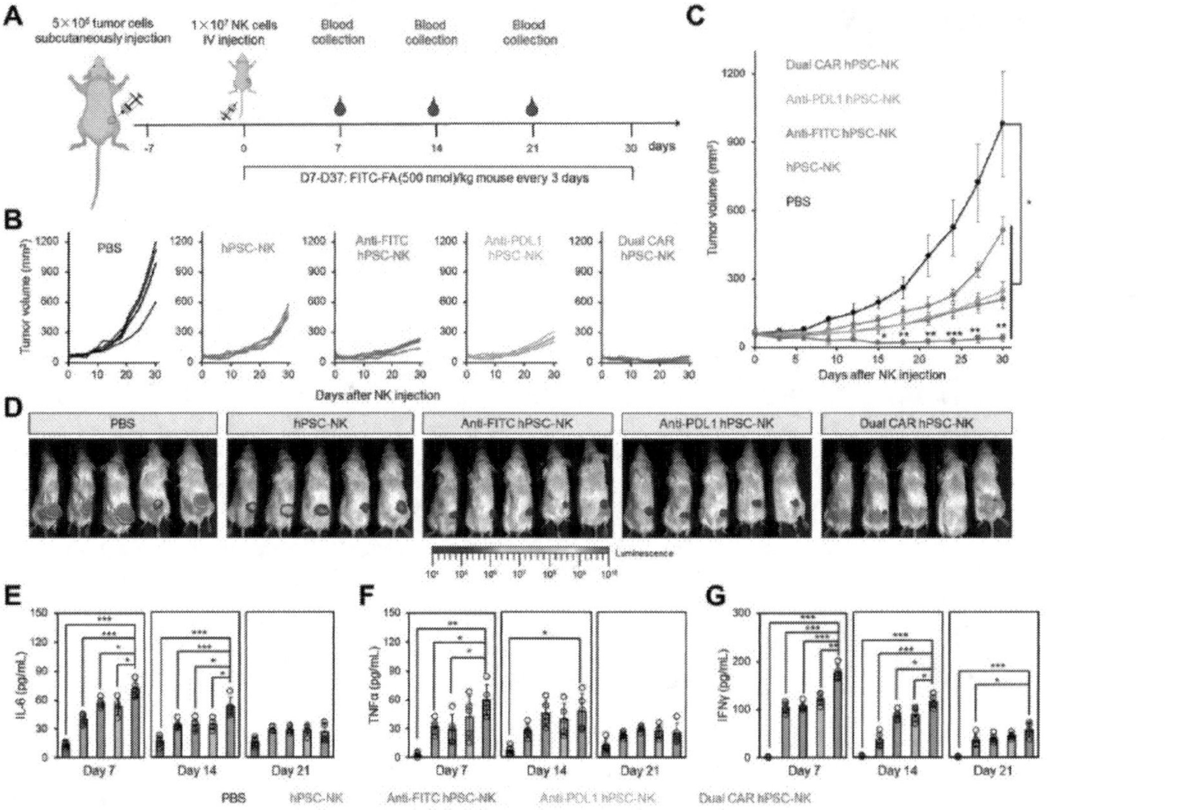

Fig. 5. *In vivo* anti-tumor activities of various hPSC-NK cells were assessed via intravenous injection. (A) Schematic of intravenous injection of various hPSC-NK cells for *in vivo* anti-tumor cytotoxicity study. 5×10^5 MDA-MB-231 cells were subcutaneously implanted into the right back of NRG mice. After 7 days, mice were intravenously treated with PBS or 1×10^7 hPSC-NK cells. (B-C) Time-dependent tumor burden was measured and quantified for indicated experimental mouse groups. $^*p < 0.05$, $^{**}p < 0.01$, and $^{***}p < 0.001$. (D) Representative photos of tumor-bearing mice at the end of treatment. Levels of released human tumor necrosis factor-α (TNFα) (E), IL-6 (F), and interferon gamma (IFNγ) (G) in peripheral blood collected from indicated experimental mice were measured by ELISA. n = 5. $^*p < 0.05$, $^{**}p < 0.01$, and $^{***}p < 0.001$.

CD28 transmembrane and FcεRIγ signaling domains [57,58], our CAR construct contains a NK-specific NKG2D transmembrane domain, and a CD3ζ signaling domain that is essential for both T and NK cell activation [59]. We also incorporated a NK-specific 2B4 co-stimulatory domain [33] into our CAR construct, and instead of scFv, we used a smaller anti-PD-L1 nanobody for tumor antigen targeting [60]. CAR-NK cells bearing an anti-FITC CAR showed significantly improved anti-tumor activity, both *in vitro* and *in vivo*, after engineering with the anti-PD-L1 CAR. While we have demonstrated the therapeutic potential of our dual CAR-NK cells in the immunodeficient mice, other preclinical models with an intact immune system, such as humanized mouse models with human immune T cells, and orthotopic tumor implantation are needed to better assess the safety and efficacy of hPSC-derived CAR-NK cells.

Primary mature NK cells present a short-lived phenotype *in vivo* [61], imparting a poor persistence in xenograft tumor models and human cancer patients [18]. Recent studies have demonstrated that mouse and human NK cells display a memory-like phenotype in response to viral infection [14,62–64], certain hapten exposure [64], and/or a brief treatment with cytokines of IL-12, IL-15 and IL-18 [10,16,20–22], leading to the increased longevity and improved effector function of NK cells in preclinical and clinical models. However, more enhanced NK cell persistence *in vivo* is still urgently needed to achieve sustained clinical benefits in various cancers, particularly in solid tumors. Systemic administration or ectopic expression of IL-15 has been widely used to improve *in vivo* persistence of various NK cells [17–19], which may lead to autonomous NK cell proliferation, severe adverse effects, or even leukemia transformation in patients [26]. IL-15 was also linked to a lymphoma/leukemia-targeting CD19 CAR via a 2A sequence peptide to improve *in vivo* persistence and anti-tumor functions of cord blood NK cells [18,65]. IL-15 activates STAT5 signaling via IL-2 receptor β-chain (IL-2Rβ), and thus we directly added a truncated IL-2Rβ (ΔIL-2Rβ) domain [45] into our tumor microenvironment responsive anti-PD-L1 CARs to achieve antigen-inducible NK cell expansion. Since IL-21 plays a critical role in memory cell formation [66], reverses NK cell exhaustion [34], and promotes expansion of memory-like NK cells [67,68], we also incorporated into our anti-PD-L1 CAR construct a STAT3 signaling activation motif YXXQ, an IL-21-associated residue within the IL-21 receptor. Such modifications have led to robust activation of STAT3 and STAT5 signaling pathway in a tumor antigen-responsive manner, and to CAR-NK cells with enhanced *in vivo* persistence and anti-tumor functions than that seen with single anti-FITC CAR NK cells lacking an anti-PD-L1 CAR. Cell expansion was not observed in CAR-NK cells with the anti-PD-L1 CAR co-culturing with PD-L1 rare tumor cells, further demonstrating the antigen specificity of our CAR-NK cell persistence for a safer and more durable anti-tumor immunity.

CAR-NK cells have been engineered from various NK cells, including NK-92 cell line, hPSC-derived, cord blood and peripheral blood NK cells, though NK-92 derived CAR-NK cells are dominant in clinical trials due to its excellent expansion capacity *in vitro*. While no obvious toxicity was observed in clinical trials with NK-92 cells [69], concerns still exist regarding their *in vivo* survival and proliferation after irradiation during cell preparation for infusion [33]. Alternatively, engineering CAR-NK cells from hPSCs has allowed an unlimited cell source of truly off-the-shelf cellular products for clinical trials [33]. In addition, the relative ease of genome editing in hPSCs also allows massive production

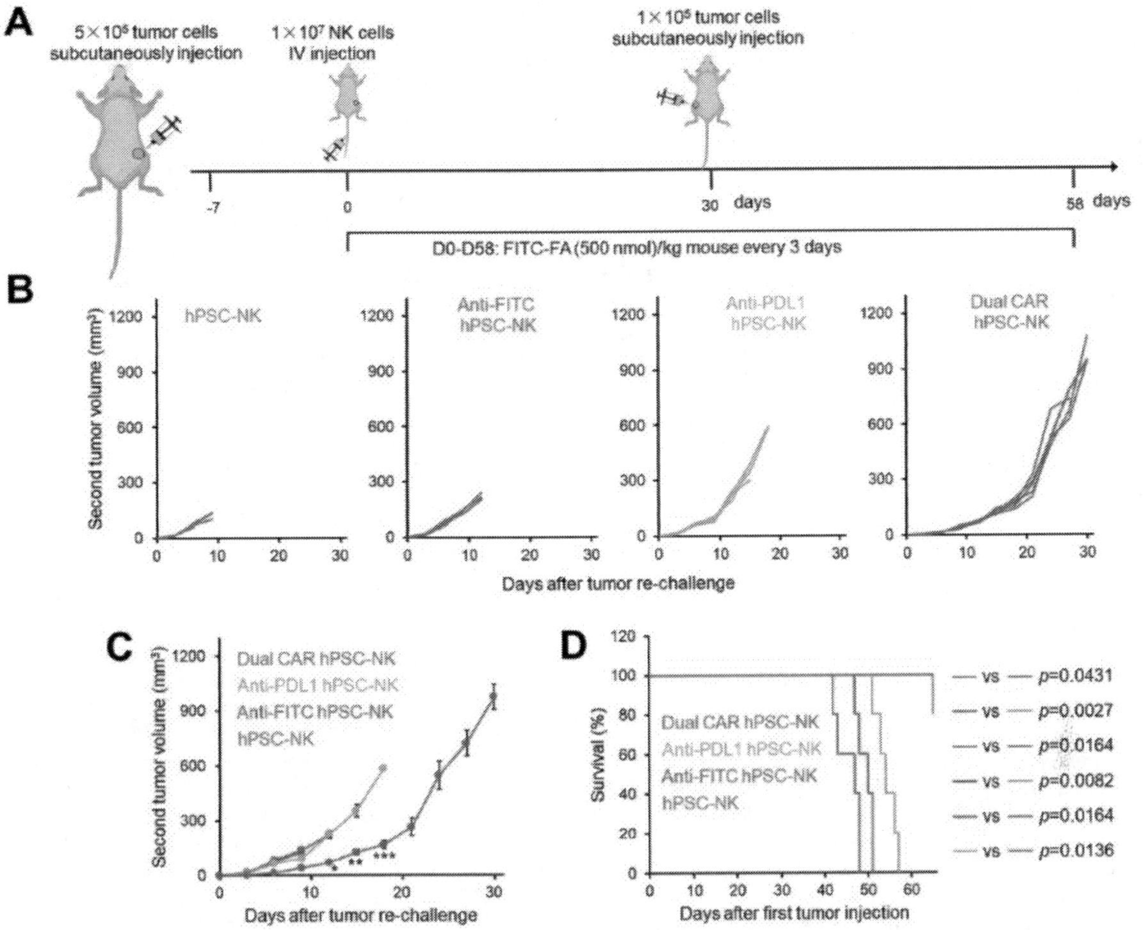

Fig. 6. Dual CAR-hPSC-NK cells have improved anti-tumor activities in an *in vivo* tumor rechallenge model. (A) Schematic of the *in vivo* tumor rechallenge model, in which 1 × 10^5 MDA-MB-231 cells were subcutaneously implanted into the left back of NSG mice at day 37. (B-C) Time-dependent second tumor burden was measured and quantified for indicated experimental mouse groups. Experimental mice were sacrificed when the primary tumor size reaches 1000 mm^3. (D) Kaplan-Meier Curve demonstrating survival of indicated experimental mouse groups was shown. n = 5.

of homogenous and stable CAR-expressing NK cells for a more standardized product on a clinical scale. In this study, we demonstrated that both CRISPR/Cas9-mediated knockin and lentiviral transduction strategies could be used to introduce CAR constructs into hPSCs for functional CAR-NK cell production.

In conclusion, we have developed a novel immunoengineering approach in which hPSCs are engineered with dual CAR constructs and differentiated into off-the-shelf CAR-NK cellular products. The dual anti-FITC and anti-PD-L1 hPSC CAR-NK cells are designed and demonstrated with improved universality, safety, potency, and persistence, both *in vitro* and *in vivo*, in an antigen-dependent manner, achieving a memory-like phenotype of NK cells.

4. Experimental section

CAR plasmid construction. To construct anti-PD-L1 lentiviral vectors, DNA sequence encoding CD8a signal peptide, anti-PD-L1 nanobody [60], CD28 extracellular domain, CD28 or NKG2D transmembrane domain, CD28 or 2B4 intracellular co-stimulatory domain, ΔIL-2Rβ, CD3ζ-YXXQ was directly synthesized and cloned into the lenti-luciferase-P2A-NeoR (Addgene #105621) backbone via NEBuilder HiFi DNA Assembly after BamHI and MluI digestion. The resulting anti-PD-L1 plasmids were then sequenced and digested with MluI to further incorporate an IRES-NeoR or IRES-GFP sequence. The anti-FITC CAR plasmid with CD8a signal peptide, anti-fluorescein scFv, CD8a extracellular and intracellular domains, 4-1BB co-stimulatory domain and CD3ζ signaling domain was previously constructed in our lab [39] and cloned into our AAVS1-Puro CAG FUCCI donor plasmid [46] (Addgene #136934). The resulting AAVS1-Puro CAG anti-FITC-CAR plasmid was digested with SgrDI and MluI, and ligated to the lentiviral anti-PD-L1 CAR backbone to construct the lentiviral anti-FITC CAR vector. To make anti-FITC CAR plasmid with NKG2D transmembrane and 2B4 co-stimulatory domains, anti-FITC scFv sequence and chimeric sequence of NKG2D, 2B4 and CD3ζ were PCR amplified from lentiviral anti-FITC CAR vector and AAVS1-Puro CAG CLTX-NKG2D-2B4-CD3z CAR (Addgene #157744), respectively, and cloned into AAVS1-Puro CAG FUCCI plasmid via NEBuilder HiFi DNA Assembly to make AAVS1-Puro CAG anti-FITC-NKG2D-2B4-CD3z CAR, which was digested with SgrDI and MluI, and ligated to the lentiviral anti-PD-L1 CAR backbone to construct the lentiviral anti-FITC-NKG2D-2B4-CD3z CAR.

Maintenance and differentiation of hPSCs. H9 hPSC line was obtained from WiCell and maintained on Matrigel-coated plates in mTeSR plus medium. For NK cell differentiation, hPSCs were dissociated with 0.5 mM EDTA and seeded onto iMatrix 511-coated 24-well plate at a cell density between 10, 000 and 80, 000 cells/cm^2 in mTesR plus medium with 5 μM Y27632 for 24 h (day −1). At day 0, cells were treated with 6 μM CHIR99021 (CHIR) in DMEM medium supplemented with 100 μg/mL ascorbic acid (DMEM/Vc), followed by a medium change with LaSR basal medium from day 1 to day 4. 50 ng/mL VEGF was added to the medium from day 2 to day 4. At day 4, medium was replaced by Stemline II medium (Sigma) supplemented with 10 μM SB431542, 25 ng/mL SCF and 25 ng/mL FLT3L. On day 6, SB431542-containing medium was aspirated and cells were maintained in Stemline II medium with 50 ng/mL SCF and 50 ng/mL FLT3L. At day 9 and day 12, top half medium was aspirated and replaced with 0.5 mL of fresh Stemline II medium containing 50 ng/mL SCF and 50 ng/mL FLT3L. At day 15, floating cells were gently harvested, filtered with a cell strainer (70 μm), and co-culture on OP9-DLL4 (kindly provided by Dr. Igor Slukvin) monolayer (2×10^4 cells/mL) in NK cell differentiation medium: α-MEM medium supplemented with 20% fetal bovine serum (FBS), 5 ng/mL IL-7, 5 ng/mL FTL3L, 25 ng/mL SCF, 5 ng/mL IL-15, and 35 nM UM171. NK cell differentiation medium was changed every 3 days, and floating cells were transferred onto fresh OP9-DLL4 monolayer every 6 days.

Lentivirus production and hPSC transduction. For lentivirus production, 293TN cells (System Biosciences) were incubated in DMEM medium containing 10% fetal bovine serum (FBS), 1% sodium pyruvate, and 0.5% GlutaMAX until 95–100% confluence. 4.5 μg lentiviral CAR plasmid, 3.0 μg psPAX2, and 1.5 μg pMD2.G were added to 450 μL of Opti-MEM medium and incubated at room temperature for 5 min. 27 μL of FuGENE HD reagent was then added to the mixture and incubated at room temperature for another 15 min. The resulting 450 μL plasmid mixture was added to 3 mL of culture medium and evenly distributed to 3 wells of 6 well plate with 293TN cells after aspirating the old medium. 18 h after plasmid addition, aspirate the medium from each well and replace with 3 mL of fresh pre-warmed culture medium, and incubated for another 24 h. Virus-containing supernatant was then collected every day with fresh warm medium change for 2–3 days, transferred to a 50 mL conical tube, and stored at 4 °C. The resulting virus supernatant were then centrifuged at 2, 000 g at 4 °C for 5 min or filtered through a 0.45 μm filter to remove cell debris. For hPSC transduction, hPSCs were dissociated with 0.5 mM EDTA and seeded onto iMatrix 511-coated 6-well plate at a cell density between 10, 000 and 80, 000 cells/cm^2 in mTesR plus medium with 5 μM Y27632. 24 h later, stem cell culture medium was aspirated and replaced with 1 mL of mTesR plus medium with 5 μM Y27632 and 1 mL of virus supernatant, which were removed and replaced with 2 mL of fresh mTeSR plus after 24 h. 2–3 days after transduction, 300 μg/ml G418 or 1 μg/mL puromycin was applied to select successfully transduced hPSCs. To further enrich desired cells, transduced hPSCs were dissociated and transferred to 96-well plate at a cell density of 10 cells/mL. After a 4-day culture, hPSCs were continuously treated with 300 μg/ml G418 or 1 μg/mL puromycin for 8 more days.

NK-92 cell maintenance and lentiviral transduction. NK-92 cells were cultured in MyeloCult H5100 medium containing 100 units/mL human recombinant IL-2. For lentiviral transduction, NK-92 cells were stimulated by IL-2 and IL-15 firstly. Briefly, count the NK-92 cells and resuspend them in appropriate medium (RPMI 1640, 10% FBS, 2 nM L-glutamine, 20 ng/mL IL-2, 50 ng/mL IL-15, and 100 ng/mL IL-12) at 1×10^6 cells/mL. These NK-92 cells were stimulated for 2 h before lentiviral transduction. After cytokine stimulation, 1×10^5 NK-92 cells were plated in each well of a 12-well plate, and cells were treated with 1 mL virus supernatant and polybrene (8 μg/mL) overnight at 37 °C, 5% CO_2. After 24 h, viruses were removed by centrifuging at 360×g for 5 min, and the resulting NK-92 cells were suspended in 1 mL MyeloCult H5100 medium with 100 units/mL human recombinant IL-2. After 5 days, transduced NK-92 cells were centrifuged at 360×g for 5 min and resuspended in 1 mL MyeloCult H5100 medium containing 100 units/mL human recombinant IL-2 and 1 μg/mL puromycin or 300 μg/ml G418. At least 8-day drug screening is needed to enrich successfully transduced NK-92 cells.

MDA-MB-231 and LNCaP cell maintenance. LNCaP tumor cells were kindly provided and cultured by the laboratory of Dr. Chang-Deng Hu at Purdue University. MDA-MB-231 cells were cultured in Leibovitz's L-15 medium (containing 10% FBS, 100 units/mL penicillin and 100 mg mL^{-1} streptomycin), and LNCaP cells were cultured in RPMI-1640 medium (containing 10% FBS, 100 units/mL penicillin and 100 mg mL^{-1} streptomycin). These two cell lines were incubated at 37 °C, 5% CO_2. The culture medium was changed every two days and cells passaged at 70–80% confluency.

Flow cytometry analysis. Differentiated cells were gently pipetted and filtered through a 70 or 100 μm strainer. The cells were then pelleted by centrifugation and washed twice with PBS −/− solution containing 1% bovine serum albumin (BSA). Cells were stained with appropriate conjugated or non-conjugated antibodies (Table S2) for 30 min at room temperature in dark and analyzed in an Accuri C6 plus cytometry (Beckton Dickinson) after washing with BSA-containing PBS −/− solution. FlowJo software was used to analyze the collected data.

NK cell-mediated *in vitro* cytotoxicity assay. The cell viability was analyzed by flow cytometry according to a previous protocol [70]. Briefly, tumor cells were stained with 2 μM Calcein-AM in MEM medium at 37 °C for 10 min in dark, followed by 10% FBS treatment for 10 min in dark at room temperature. Labeled tumor cells were pelleted at 300×g for 7 min and resuspended in culture medium with 10% FBS at a density of 50,000 cells/mL. 100 μL of tumor cells were then mixed with 100 μL of 150,000, 250,000, and 500,000 cells/mL NK cells in 96 well plates with or without FITC-folate (10 nmol/L), and incubated at 37 °C, 5% CO_2 for 12 h. To harvest all the cells, cell-containing medium was firstly transferred into a new round-bottom 96-well plate, and 50 μL trypsin-EDTA was added to the empty wells to dissociate attached cells. After 5 min incubation at 37 °C, dissociated cells were transferred into the same wells of round-bottom 96-well plate with floating cells. All cells were pelleted by centrifuging (300×g, 4 °C, 5 min), and washed with 200 μL of PBS−/− solution containing 0.5% BSA. The pelleted cells were stained with propidium iodide (PI) for 15 min at room temperature, and analyzed in the Accuri C6 plus cytometer (Beckton Dickinson).

Conjugate formation assay. To visualize immunological synapses, 100 μL of tumor cells (50,000 cells/mL) were seeded onto wells of 96-well plate and incubated at 37 °C for 12 h, allowing them to attach. 100 μL NK cells (500,000 cells/mL) were then added onto the target tumor cells and incubated for 6 h before fixing with 4% paraformaldehyde (in PBS). Cytoskeleton staining was then performed using an F-actin Visualization Biochem Kit (Cytoskeleton Inc.).

ELISA analysis. To analyze the cytokine production by ELISA assay, 100 μL of tumor cells (50,000 cells/mL) were seeded onto wells of 96-well plate and incubated at 37 °C for 12 h, allowing them to attach. 100 μL NK cells (500,000 cells/mL) with or without FITC-folate (10 nmol/L) were then added onto the target tumor cells and incubated for 6 h. Afterwards, plates were centrifuged at 350×g for 10 min to spin down the cell debris, and 10 μL top supernatant was collected for measuring the TNFα and IL-6 production using an ELISA kit (ThermoFisher Scientific, US).

MDA-MG-231 xenograft studies. All the mouse experiments were approved by the Purdue Animal Care and Use Committee (PACUC). The immunodeficient NOD.Cg-$RAG^{1tm1Mom}IL2rg^{tm1Wjl}$/SzJ (NRG) mice were bred and maintained by the Biological Evaluation Core at the Purdue University Center for Cancer Research. MDA-MB-231 cells (5×10^5 tumor cells/per mouse) were implanted subcutaneously. When the tumor size reached ~100 mm^3, NK cells and FITC-folate were intravenously injected. Mice were maintained on folic acid-deficient diet (TD.95247, Envigo) in order to reduce the level of folic acid in mice to a physiological level found in humans. Tumors were measured the other day with caliper, and tumor volume was calculated according to the equation: tumor volume $= L \times W^2 \times 1/2$, where L is the longest axis of the tumor and W is the axis perpendicular to L. Mouse blood was also collected for NK cell and cytokine release (TNFα and IL-6) analysis, and systemic toxicity was monitored by measuring body weight loss of experimental mice.

Statistical analysis. Data are presented as mean ± standard deviation (SD). Statistical significance was determined by Student's *t*-test (two-tail) between two groups, and three or more groups were analyzed by one-way analysis of variance (ANOVA). $P < 0.05$ was considered statistically significant.

Ethics approval and consent to participate

No human studies were involved in this project. All the mouse experiments were approved by the Purdue Animal Care and Use Committee (PACUC).

CRediT authorship contribution statement

Yun Chang: Conceptualization, Formal analysis, Investigation, Writing – original draft. **Gyuhyung Jin:** Validation, Writing – review & editing. **Weichuan Luo:** Resources, Validation, Writing – review & editing. **Qian Luo:** Resources, Validation, Writing – review & editing. **Juhyung Jung:** Validation, Writing – review & editing. **Sydney N. Hummel:** Validation, Writing – review & editing. **Sandra Torregrosa-Allen:** Resources, Methodology, Validation, Writing – review & editing. **Bennett D. Elzey:** Resources, Methodology, Validation, Supervision, Writing – review & editing. **Philip S. Low:** Resources, Supervision, Writing – review & editing. **Xiaojun Lance Lian:** Resources, Supervision, Funding acquisition, Writing – review & editing. **Xiaoping Bao:** Supervision, Funding acquisition, Writing – review & editing.

Declaration of competing interest

A patent related to this manuscript is under application (Y.C., J.J., P.S.L., and X.B.).

Acknowledgements

We thank members of the Low and Bao laboratories for technical assistance. We're also gratefully acknowledge the Purdue Flow Cytometry and Cell Separation Facility, Purdue Genomics Core Facility and the Biological Evaluation Core at Purdue University Center for Cancer Research (PCCR). This study was supported by startup funding from the Davidson School of Chemical Engineering and the College of Engineering at Purdue (X.B.), PICR Robbers New Investigators (X.B.), Showalter Research Trust (Young Investigator Award to X.B.), NSF CBET (grant no. 2143064 to X.B.), NSF CBET (grant no. 1943696 to X.L.L.), and NIH NCI (grant no. R37CA265926 to X.B.). The authors also gratefully acknowledge support from the Purdue University Institute for Cancer Research, P30CA023168, Purdue Institute for Integrative Neuroscience (PIIN) and Bindley Biosciences Center, and the Walther Cancer Foundation.

Appendix A. Supplementary data

Supplementary data to this article can be found online at https://doi.org/10.1016/j.bioactmat.2023.03.018.

References

[1] M.W. Rohaan, S. Wilgenhof, J.B.A.G. Haanen, Adoptive cellular therapies: the current landscape, Virchows Arch. (2019), https://doi.org/10.1007/s00428-018-2484-0.
[2] C.S. Hinrichs, S.A. Rosenberg, Exploiting the curative potential of adoptive T-cell therapy for cancer, Immunol. Rev. (2014), https://doi.org/10.1111/imr.12132.
[3] A. Redeker, R. Arens, Improving adoptive T cell therapy: the particular role of T cell costimulation, cytokines, and post-transfer vaccination, Front. Immunol. (2016), https://doi.org/10.3389/fimmu.2016.00345.
[4] H. Zhu, R.H. Blum, D. Bernareggi, E.H. Ask, Z. Wu, H.J. Hoel, Z. Meng, C. Wu, K. L. Guan, K.J. Malmberg, D.S. Kaufman, Metabolic reprograming via deletion of CISH in human iPSC-derived NK cells promotes in vivo persistence and enhances anti-tumor activity, Cell Stem Cell (2020), https://doi.org/10.1016/j.stem.2020.05.008.
[5] B.H. Goldenson, H. Zhu, Y.Z.M. Wang, N. Heragu, D. Bernareggi, A. Ruiz-Cisneros, A. Bahena, E.H. Ask, H.J. Hoel, K.J. Malmberg, D.S. Kaufman, Umbilical cord blood and iPSC-derived natural killer cells demonstrate key differences in cytotoxic activity and KIR profiles, Front. Immunol. (2020), https://doi.org/10.3389/fimmu.2020.561553.
[6] F. Cichocki, R. Bjordahl, S. Gaidarova, S. Mahmood, R. Abujarour, H. Wang, K. Tuininga, M. Felices, Z.B. Davis, L. Bendzick, R. Clarke, L. Stokely, P. Rogers, M. Ge, M. Robinson, B. Rezner, D.L. Robbins, T.T. Lee, D.S. Kaufman, B.R. Blazar, B. Valamehr, J.S. Miller, iPSC-derived NK cells maintain high cytotoxicity and enhance in vivo tumor control in concert with T cells and anti–PD-1 therapy, Sci. Transl. Med. (2020), https://doi.org/10.1126/scitranslmed.aaz5618.
[7] H. Zhu, R.H. Blum, R. Bjordahl, S. Gaidarova, P. Rogers, T.T. Lee, R. Abujarour, G. B. Bonello, J. Wu, P.F. Tsai, J.S. Miller, B. Walcheck, B. Valamehr, D.S. Kaufman, Pluripotent stem cell-derived NK cells with high-affinity noncleavable CD16a mediate improved antitumor activity, Blood (2020), https://doi.org/10.1182/blood.2019000621.
[8] Y. Chang, X. Bao, Adoptive natural killer cell therapy: a human pluripotent stem cell perspective, Curr. Opin. Chem. Eng. (2020), https://doi.org/10.1016/j.coche.2020.08.008.
[9] R. Handgretinger, P. Lang, M.C. André, Exploitation of natural killer cells for the treatment of acute leukemia, Blood (2016), https://doi.org/10.1182/blood-2015-12-629055.
[10] R. Romee, M. Rosario, M.M. Berrien-Elliott, J.A. Wagner, B.A. Jewell, T. Schappe, J.W. Leong, S. Abdel-Latif, S.E. Schneider, S. Willey, C.C. Neal, L. Yu, S.T. Oh, Y. S. Lee, A. Mulder, F. Claas, M.A. Cooper, T.A. Fehniger, Cytokine-induced memory-like natural killer cells exhibit enhanced responses against myeloid leukemia, Sci. Transl. Med. (2016), https://doi.org/10.1126/scitranslmed.aaf2341.
[11] A. Cerwenka, L.L. Lanier, Natural killer cell memory in infection, inflammation and cancer, Nat. Rev. Immunol. (2016), https://doi.org/10.1038/nri.2015.9.
[12] S. Paust, U.H. Von Andrian, Natural killer cell memory, Nat. Immunol. (2011), https://doi.org/10.1038/ni.2032.
[13] T.E. O'Sullivan, J.C. Sun, L.L. Lanier, Natural killer cell memory, Immunity (2015), https://doi.org/10.1016/j.immuni.2015.09.013.
[14] J.C. Sun, J.N. Beilke, L.L. Lanier, Adaptive immune features of natural killer cells, Nature (2009), https://doi.org/10.1038/nature07665.
[15] M.M. Berrien-Elliott, J.A. Wagner, T.A. Fehniger, Human cytokine-induced memory-like natural killer cells, J. Innate Immun. (2015), https://doi.org/10.1159/000382019.
[16] M.A. Cooper, J.M. Elliott, P.A. Keyel, L. Yang, J.A. Carrero, W.M. Yokoyama, Cytokine-induced memory-like natural killer cells, Proc. Natl. Acad. Sci. U.S.A. (2009), https://doi.org/10.1073/pnas.0813192106.
[17] R. Ma, T. Lu, Z. Li, K.Y. Teng, A.G. Mansour, M. Yu, L. Tian, B. Xu, S. Ma, J. Zhang, T. Barr, Y. Peng, M.A. Caligiuri, J. Yu, An Oncolytic Virus Expressing Il15/il15ra Combined with Off-The-Shelf Egfr-Car Nk Cells Targets Glioblastoma, Cancer Res., 2021, https://doi.org/10.1158/0008-5472.CAN-21-0035.
[18] E. Liu, Y. Tong, G. Dotti, H. Shaim, B. Savoldo, M. Mukherjee, J. Orange, X. Wan, X. Lu, A. Reynolds, M. Gagea, P. Banerjee, R. Cai, M.H. Bdaiwi, R. Basar, M. Muftuoglu, L. Li, D. Marin, W. Wierda, M. Keating, R. Champlin, E. Shpall, K. Rezvani, Cord blood NK cells engineered to express IL-15 and a CD19-targeted CAR show long-term persistence and potent antitumor activity, Leukemia (2018), https://doi.org/10.1038/leu.2017.226.
[19] Z. Du, Y.Y. Ng, S. Zha, S. Wang, piggyBac system to co-express NKG2D CAR and IL-15 to augment the in vivo persistence and anti-AML activity of human peripheral blood NK cells, Mol. Ther. - Methods Clin. Dev. (2021), https://doi.org/10.1016/j.omtm.2021.10.014.
[20] R. Romee, S.E. Schneider, J.W. Leong, J.M. Chase, C.R. Keppel, R.P. Sullivan, M. A. Cooper, T.A. Fehniger, Cytokine activation induces human memory-like NK cells, Blood. https://doi.org/10.1182/blood-2012-04-419283, 2012.
[21] M. Gang, N.D. Marin, P. Wong, C.C. Neal, L. Marsala, M. Foster, T. Schappe, W. Meng, J. Tran, M. Schaettler, M. Davila, F. Gao, A.F. Cashen, N.L. Bartlett, N. Mehta-Shah, B.S. Kahl, M.Y. Kim, M.L. Cooper, J.F. DiPersio, M.M. Berrien-Elliott, T.A. Fehniger, CAR-modified memory-like NK cells exhibit potent responses to NK-resistant lymphomas, Blood (2020), https://doi.org/10.1182/BLOOD.2020006619.
[22] S. Lopez-Vergès, J.M. Milush, S. Pandey, V.A. York, J. Arakawa-Hoyt, H. Pircher, P. J. Norris, D.F. Nixon, L.L. Lanier, CD57 defines a functionally distinct population of mature NK cells in the human CD56dimCD16+ NK-cell subset, Blood (2010), https://doi.org/10.1182/blood-2010-04-282301.
[23] K. Skak, K.S. Frederiksen, D. Lundsgaard, Interleukin-21 activates human natural killer cells and modulates their surface receptor expression, Immunology (2008), https://doi.org/10.1111/j.1365-2567.2007.02730.x.
[24] A. Heinze, B. Grebe, M. Bremm, S. Huenecke, T.A. Munir, L. Graafen, J.T. Frueh, M. Merker, E. Rettinger, J. Soerensen, T. Klingebiel, P. Bader, E. Ullrich, C. Cappel, The synergistic use of IL-15 and IL-21 for the generation of NK cells from CD3/CD19-depleted grafts improves their ex vivo expansion and cytotoxic potential against neuroblastoma: perspective for optimized immunotherapy post haploidentical stem cell trans, Front. Immunol. (2019), https://doi.org/10.3389/fimmu.2019.02816.
[25] M. Ma, S. Badeti, J.K. Kim, D. Liu, Natural killer (NK) and CAR-NK cell expansion method using membrane bound-IL-21-modified B cell line, J. Vis. Exp. (2022), https://doi.org/10.3791/62336.
[26] A. Mishra, S. Liu, G.H. Sams, D.P. Curphey, R. Santhanam, L.J. Rush, D. Schaefer, L.G. Falkenberg, L. Sullivan, L. Jaroncyk, X. Yang, H. Fisk, L.C. Wu, C. Hickey, J. C. Chandler, Y.Z. Wu, N.A. Heerema, K.K. Chan, D. Perrotti, J. Zhang, P. Porcu, F. K. Racke, R. Garzon, R.J. Lee, G. Marcucci, M.A. Caligiuri, Aberrant Overexpression of IL-15 Initiates Large Granular Lymphocyte Leukemia through Chromosomal Instability and DNA Hypermethylation, Cancer Cell, 2012, https://doi.org/10.1016/j.ccr.2012.09.009.
[27] M. Liu, W. Song, L. Huang, Drug Delivery Systems Targeting Tumor-Associated Fibroblasts for Cancer Immunotherapy, Cancer Lett, 2019, https://doi.org/10.1016/j.canlet.2019.01.032.

[28] W. Song, L. Shen, Y. Wang, Q. Liu, T.J. Goodwin, J. Li, O. Dorosheva, T. Liu, R. Liu, L. Huang, Synergistic and low adverse effect cancer immunotherapy by immunogenic chemotherapy and locally expressed PD-L1 trap, Nat. Commun. (2018), https://doi.org/10.1038/s41467-018-04605-x.

[29] H. Xu, M. Hu, M. Liu, S. An, K. Guan, M. Wang, L. Li, J. Zhang, J. Li, L. Huang, Nano-puerarin Regulates Tumor Microenvironment and Facilitates Chemo- and Immunotherapy in Murine Triple Negative Breast Cancer Model, Biomaterials, 2020, https://doi.org/10.1016/j.biomaterials.2020.119769.

[30] M. Liu, A.R. Khan, J. Ji, G. Lin, X. Zhao, G. Zhai, Crosslinked self-assembled nanoparticles for chemo-sonodynamic combination therapy favoring antitumor, antimetastasis management and immune responses, J. Contr. Release (2018), https://doi.org/10.1016/j.jconrel.2018.10.007.

[31] G. Chen, A.C. Huang, W. Zhang, G. Zhang, M. Wu, W. Xu, Z. Yu, J. Yang, B. Wang, H. Sun, H. Xia, Q. Man, W. Zhong, L.F. Antelo, B. Wu, X. Xiong, X. Liu, L. Guan, T. Li, S. Liu, R. Yang, Y. Lu, L. Dong, S. McGettigan, R. Somasundaram, R. Radhakrishnan, G. Mills, Y. Lu, J. Kim, Y.H. Chen, H. Dong, Y. Zhao, G. C. Karakousis, T.C. Mitchell, L.M. Schuchter, M. Herlyn, E.J. Wherry, X. Xu, W. Guo, Exosomal PD-L1 contributes to immunosuppression and is associated with anti-PD-1 response, Nature (2018), https://doi.org/10.1038/s41586-018-0392-8.

[32] X. Jiang, J. Wang, X. Deng, F. Xiong, J. Ge, B. Xiang, X. Wu, J. Ma, M. Zhou, X. Li, Y. Li, G. Li, W. Xiong, C. Guo, Z. Zeng, Role of the tumor microenvironment in PD-L1/PD-1-mediated tumor immune escape, Mol. Cancer (2019), https://doi.org/10.1186/s12943-018-0928-4.

[33] Y. Li, D.L. Hermanson, B.S. Moriarity, D.S. Kaufman, Human iPSC-derived natural killer cells engineered with chimeric antigen receptors enhance anti-tumor activity, Cell Stem Cell (2018), https://doi.org/10.1016/j.stem.2018.06.002.

[34] H. Seo, I. Jeon, B.S. Kim, M. Park, E.A. Bae, B. Song, C.H. Koh, K.S. Shin, I.K. Kim, K. Choi, T. Oh, J. Min, B.S. Min, Y.D. Han, S.J. Kang, S.J. Shin, Y. Chung, C.Y. Kang, IL-21-mediated reversal of NK cell exhaustion facilitates anti-Tumour immunity in MHC class I-deficient tumours, Nat. Commun. (2017), https://doi.org/10.1038/ncomms15776.

[35] S.J. Judge, W.J. Murphy, R.J. Canter, Characterizing the dysfunctional NK cell: assessing the clinical relevance of exhaustion, anergy, and senescence, Front. Cell. Infect. Microbiol. (2020), https://doi.org/10.3389/fcimb.2020.00049.

[36] Q. Luo, J.V. Napoleon, X. Liu, B. Zhang, S. Zheng, P.S. Low, Targeted rejuvenation of exhausted chimeric antigen receptor T-cells regresses refractory solid tumors, Mol. Cancer Res. (2022), https://doi.org/10.1158/1541-7786.mcr-21-0711.

[37] K. Tamada, D. Geng, Y. Sakoda, N. Bansal, R. Srivastava, Z. Li, E. Davila, Redirecting gene-modified T cells toward various cancer types using tagged antibodies, Clin. Cancer Res. (2012), https://doi.org/10.1158/1078-0432.CCR-12-1449.

[38] J.S.Y. Ma, J.Y. Kim, S.A. Kazane, S.H. Choi, H.Y. Yun, M.S. Kim, D.T. Rodgers, H. M. Pugh, O. Singer, S.B. Sun, B.R. Fonslow, J.N. Kochenderfer, T.M. Wright, P. G. Schultz, T.S. Young, C.H. Kim, Y. Cao, Versatile strategy for controlling the specificity and activity of engineered T cells, Proc. Natl. Acad. Sci. U.S.A. (2016), https://doi.org/10.1073/pnas.1524193113.

[39] Y.G. Lee, H. Chu, Y. Lu, C.P. Leamon, M. Srinivasarao, K.S. Putt, P.S. Low, Regulation of CAR T cell-mediated cytokine release syndrome-like toxicity using low molecular weight adapters, Nat. Commun. (2019), https://doi.org/10.1038/s41467-019-10565-7.

[40] Y.G. Lee, I. Marks, M. Srinivasarao, A.K. Kanduluru, S.M. Mahalingam, X. Liu, H. Chu, P.S. Low, Use of a Single CAR T Cell and Several Bispecific Adapters Facilitates Eradication of Multiple Antigenically Different Solid Tumors, Cancer Res., 2019, https://doi.org/10.1158/0008-5472.CAN-18-1834.

[41] H. Zhu, Y.S. Lai, Y. Li, R.H. Blum, D.S. Kaufman, Concise review: human pluripotent stem cells to produce cell-based cancer immunotherapy, Stem Cell. (2018), https://doi.org/10.1002/stem.2754.

[42] M. Carlsten, R.W. Childs, Genetic manipulation of NK cells for cancer immunotherapy: techniques and clinical implications, Front. Immunol. (2015), https://doi.org/10.3389/fimmu.2015.00266.

[43] D.A. Knorr, Z. Ni, D. Hermanson, M.K. Hexum, L. Bendzick, L.J.N. Cooper, D. A. Lee, D.S. Kaufman, Clinical-scale derivation of natural killer cells from human pluripotent stem cells for cancer therapy, Stem Cells Transl. Med (2013), https://doi.org/10.5966/sctm.2012-0084.

[44] A.M. Martin, T.R. Nirschl, C.J. Nirschl, B.J. Francica, C.M. Kochel, A. Van Bokhoven, A.K. Meeker, M.S. Lucia, R.A. Anders, A.M. Demarzo, C.G. Drake, Paucity of PD-L1 Expression in Prostate Cancer: Innate and Adaptive Immune Resistance, Prostate Cancer Prostatic Dis, 2015, https://doi.org/10.1038/pcan.2015.39.

[45] Y. Kagoya, S. Tanaka, T. Guo, M. Anczurowski, C.H. Wang, K. Saso, M.O. Butler, M. D. Minden, N. Hirano, A novel chimeric antigen receptor containing a JAK-STAT signaling domain mediates superior antitumor effects, Nat. Med. (2018), https://doi.org/10.1038/nm.4478.

[46] Y. Chang, P.B. Hellwarth, L.N. Randolph, Y. Sun, Y. Xing, W. Zhu, X.L. Lian, X. Bao, Fluorescent indicators for continuous and lineage-specific reporting of cell-cycle phases in human pluripotent stem cells, Biotechnol. Bioeng. (2020), https://doi.org/10.1002/bit.27352.

[47] Y. Chang, R. Syahirah, S.N. Oprescu, X. Wang, J. Jung, S.H. Cooper, S. Torregrosa-Allen, B.D. Elzey, A.Y. Hsu, L.N. Randolph, Y. Sun, S. Kuang, H.E. Broxmeyer, Q. Deng, X. Lian, X. Bao, Chemically-defined generation of human hemogenic endothelium and definitive hematopoietic progenitor cells, Biomaterials 285 (2022), 121569, https://doi.org/10.1016/J.BIOMATERIALS.2022.121569.

[48] Y. Chang, R. Syahirah, X. Wang, S. Toregrosa-Allen, B.D. Elzey, X. Lian, Q. Deng, H. E. Broxmeyer, X. Bao, RESOURCE Engineering Chimeric Antigen Receptor Neutrophils from Human Pluripotent Stem Cells for Targeted Cancer Immunotherapy Running Title: Generation of CAR-Neutrophils from hPSCs, 2022.

[49] G. Roex, D. Campillo-Davo, D. Flumens, P.A.G. Shaw, L. Krekelbergh, H. De Reu, Z. N. Berneman, E. Lion, S. Anguille, Two for one: targeting BCMA and CD19 in B-cell malignancies with off-the-shelf dual-CAR NK-92 cells, J. Transl. Med. (2022), https://doi.org/10.1186/s12967-022-03326-6.

[50] T. Ingegnere, F.R. Mariotti, A. Pelosi, C. Quintarelli, B. De Angelis, N. Tumino, F. Besi, C. Cantoni, F. Locatelli, P. Vacca, L. Moretta, Human CAR NK cells: a new non-viral method allowing high efficient transfection and strong tumor cell killing, Front. Immunol. (2019), https://doi.org/10.3389/fimmu.2019.00957.

[51] E. Liu, D. Marin, P. Banerjee, H.A. Macapinlac, P. Thompson, R. Basar, L. Nassif Kerbauy, B. Overman, P. Thall, M. Kaplan, V. Nandivada, I. Kaur, A. Nunez Cortes, K. Cao, M. Daher, C. Hosing, E.N. Cohen, P. Kebriaei, R. Mehta, S. Neelapu, Y. Nieto, M. Wang, W. Wierda, M. Keating, R. Champlin, E.J. Shpall, K. Rezvani, Use of CAR-transduced natural killer cells in CD19-positive lymphoid tumors, N. Engl. J. Med. (2020), https://doi.org/10.1056/nejmoa1910607.

[52] Y. Liu, Y. Zhou, K.H. Huang, X. Fang, Y. Li, F. Wang, L. An, Q. Chen, Y. Zhang, A. Shi, S. Yu, J. Zhang, Targeting epidermal growth factor-overexpressing triple-negative breast cancer by natural killer cells expressing a specific chimeric antigen receptor, Cell Prolif. (2020), https://doi.org/10.1111/cpr.12858.

[53] X. Chen, J. Han, J. Chu, L. Zhang, J. Zhang, C. Chen, L. Chen, Y. Wang, H. Wang, L. Yi, J.B. Elder, Q.E. Wang, X. He, B. Kaur, E.A. Chiocca, J. Yu, A combinational therapy of EGFR-CAR NK cells and oncolytic herpes simplex virus 1 for breast cancer brain metastases, Oncotarget (2016), https://doi.org/10.18632/oncotarget.8526.

[54] C. Zhang, M.C. Burger, L. Jennewein, S. Genßler, K. Schönfeld, P. Zeiner, E. Hattingen, P.N. Harter, M. Mittelbronn, T. Tonn, J.P. Steinbach, W.S. Wels, ErbB2/HER2-Specific NK cells for targeted therapy of glioblastoma, J. Natl. Cancer Inst. (2016), https://doi.org/10.1093/jnci/djv375.

[55] I.M. Montagner, A. Penna, G. Fracasso, D. Carpanese, A. Dalla Pietà, V. Barbieri, G. Zuccolotto, A. Rosato, Anti-PSMA CAR-engineered NK-92 cells: an off-the-shelf cell therapy for prostate cancer, Cells (2020), https://doi.org/10.3390/cells9061382.

[56] A. Nellan, D.W. Lee, Paving the road ahead for CD19 CAR T-cell therapy, Curr. Opin. Hematol. (2015), https://doi.org/10.1097/MOH.0000000000000182.

[57] Y. Robbins, S. Greene, J. Friedman, P.E. Clavijo, C. Van Waes, K.P. Fabian, M. R. Padget, H.A. Sater, J.H. Lee, P. Soon-Shiong, J. Gulley, J. Schlom, J.W. Hodge, C.T. Allen, Tumor Control via Targeting Pd-L1 with Chimeric Antigen Receptor Modified Nk Cells, Elife, 2020, https://doi.org/10.7554/eLife.54854.

[58] K.P. Fabian, M.R. Padget, R.N. Donahue, K. Solocinski, Y. Robbins, C.T. Allen, J. H. Lee, S. Rabizadeh, P. Soon-Shiong, J. Schlom, J.W. Hodge, PD-L1 targeting high-affinity NK (t-haNK) cells induce direct antitumor effects and target suppressive MDSC populations, J. Immunother. Cancer. (2020), https://doi.org/10.1136/jitc-2019-000450.

[59] S.D. Reighard, S.A. Cranert, K.M. Rangel, A. Ali, I.E. Gyurova, A.T. de la Cruz-Lynch, J.A. Tuazon, M.V. Khodoun, L.C. Kottyan, D.F. Smith, H.I. Brunner, S. N. Waggoner, Therapeutic Targeting of Follicular T Cells with Chimeric Antigen Receptor-Expressing Natural Killer Cells, Cell Reports Med, 2020, https://doi.org/10.1016/j.xcrm.2020.100003.

[60] F. Zhang, H. Wei, X. Wang, Y. Bai, P. Wang, J. Wu, X. Jiang, Y. Wang, H. Cai, T. Xu, A. Zhou, Structural basis of a novel PD-L1 nanobody for immune checkpoint blockade, Cell Discov (2017), https://doi.org/10.1038/celldisc.2017.4.

[61] Y. Zhang, D.L. Wallace, C.M. De Lara, H. Ghattas, B. Asquith, A. Worth, G. E. Griffin, G.P. Taylor, D.F. Tough, P.C.L. Beverley, D.C. Macallan, In vivo kinetics of human natural killer cells: the effects of ageing and acute and chronic viral infection, Immunology (2007), https://doi.org/10.1111/j.1365-2567.2007.02573.x.

[62] J.C. Sun, J.N. Beilke, N.A. Bezman, L.L. Lanier, Homeostatic proliferation generates long-lived natural killer cells that respond against viral infection, J. Exp. Med. (2011), https://doi.org/10.1084/jem.20100479.

[63] B. Foley, S. Cooley, M.R. Verneris, J. Curtsinger, X. Luo, E.K. Waller, C. Anasetti, D. Weisdorf, J.S. Miller, Human cytomegalovirus (CMV)-Induced memory-like NKG2C + NK cells are transplantable and expand in vivo in response to recipient CMV antigen, J. Immunol. (2012), https://doi.org/10.4049/jimmunol.1201964.

[64] S. Paust, H.S. Gill, B.Z. Wang, M.P. Flynn, E.A. Moseman, B. Senman, M. Szczepanik, A. Telenti, P.W. Askenase, R.W. Compans, U.H. Von Andrian, Critical role for the chemokine receptor CXCR6 in NK cell-mediated antigen-specific memory of haptens and viruses, Nat. Immunol. (2010), https://doi.org/10.1038/ni.1953.

[65] V. Hoyos, B. Savoldo, C. Quintarelli, A. Mahendravada, M. Zhang, J. Vera, H. E. Heslop, C.M. Rooney, M.K. Brenner, G. Dotti, Engineering CD19-specific T lymphocytes with interleukin-15 and a suicide gene to enhance their anti-lymphoma/leukemia effects and safety, Leukemia (2010), https://doi.org/10.1038/leu.2010.75.

[66] M.T. Kasaian, M.J. Whitters, L.L. Carter, L.D. Lowe, J.M. Jussif, B. Deng, K. A. Johnson, J.A.S. Witek, M. Senices, R.F. Konz, A.L. Wurster, D.D. Donaldson, M. Collins, D.A. Young, M.J. Grusby, IL-21 limits NK cell responses and promotes antigen-specific T cell activation: a mediator of the transition from innate to adaptive immunity, Immunity (2002), https://doi.org/10.1016/S1074-7613(02)00295-9.

[67] S. Venkatasubramanian, S. Cheekatla, P. Paidipally, D. Tripathi, E. Welch, A. R. Tvinnereim, R. Nurieva, R. Vankayalapati, IL-21-dependent expansion of memory-like NK cells enhances protective immune responses against Mycobacterium tuberculosis, Mucosal Immunol. (2017), https://doi.org/10.1038/mi.2016.105.

[68] M. Granzin, A. Stojanovic, M. Miller, R. Childs, V. Huppert, A. Cerwenka, Highly efficient IL-21 and feeder cell-driven ex vivo expansion of human NK cells with

therapeutic activity in a xenograft mouse model of melanoma, OncoImmunology (2016), https://doi.org/10.1080/2162402X.2016.1219007.

[69] A. Biederstädt, K. Rezvani, Engineering the next generation of CAR-NK immunotherapies, Int. J. Hematol. (2021), https://doi.org/10.1007/s12185-021-03209-4.

[70] F. Kandarian, G.M. Sunga, D. Arango-Saenz, M. Rossetti, A flow cytometry-based cytotoxicity assay for the assessment of human NK cell activity, J. Vis. Exp. 2017 (2017), https://doi.org/10.3791/56191.

154d Short and Long-Term Vascular Network Patterning Using Hybrid Microsurgery and Granular Hydrogel Scaffolds

Zaman Ataie, Jessica El-Mallah, Summer Horchler, Dino Ravnic, Amir Sheikhi

In reconstructive surgery, hydrogels are often used for tissue repair and regeneration. However, conventional nanoporous bulk hydrogels have limitations, particularly slow and random vascularization, often resulting in impaired tissue healing. To overcome these challenges, this study presents a novel strategy that combines microsurgical techniques with granular hydrogel scaffolds (GHS) for enhanced neovascularization. Different sizes of gelatin methacryloyl (GelMA) droplets were transformed into microscale hydrogel particles (microgels) by physical crosslinking, followed by chemical assembly to yield GHS with varying pore sizes. Micropuncture (MP) was performed in a rat hindlimb model via perforating blood vessels, followed by implanting the GHS on the blood vessels. Immunostaining and angiography confirmed accelerated and guided microvascular network formation using MP/GHS compared with the MP/bulk hydrogel counterpart. Furthermore, we investigated the extent of vascularization and macrophage accumulation after 7 days and compared them with a later time point (28 days). Our results showed increased vascular loops, branches, and density 28-days after surgery. Additionally, tube length and intercapillary distance remain relatively unchanged after 4 weeks of implantation. Overall, our findings suggest that this hybrid approach may hold promise for accelerating and guiding microvascular network formation, which may open new avenues for reconstructive surgery and regenerative engineering.

AUTHOR INDEX

AIChE
120 Wall Street, FL 23
New York, NY 10005-4020

ISBN 979-8-3313-1684-6

9 798331 316846

CURRAN ASSOCIATES INC.
proceedings.com